THE MEDIEVAL WARM PERIOD

Edited by

MALCOLM K. HUGHES
Laboratory of Tree Ring Research,
University of Arizona, Tucson, Arizona, U.S.A.

and

HENRY F. DIAZ
NOAA/ERL/CDC, Boulder, Colorado, U.S.A.

Reprinted from Climatic Change
Vol. 26, Nos. 2–3 (1994)

SPRINGER-SCIENCE+BUSINESS MEDIA, B.V.

A C.I.P. Catalogue record for this book is available from the Library of Congress.

ISBN 978-94-010-4518-6   ISBN 978-94-011-1186-7 (eBook)
DOI 10.1007/978-94-011-1186-7

Printed on acid-free paper

Contents

These papers result from a workshop held in Tucson, Arizona, November 5–8, 1991, with support from the NOAA Paleoclimatology Program, the Office of Energy Research of the U.S. Department of Energy, and the Climate Dynamics Program of the National Science Foundation.

WAS THERE A 'MEDIEVAL WARM PERIOD', AND IF SO, WHERE AND WHEN?

MALCOLM K. HUGHES

Laboratory of Tree Ring Research, University of Arizona, Tucson, AZ 85721 (address for correspondence), and Cooperative Institute for Research in Environmental Science, University of Colorado, Boulder, CO 80309, U.S.A.

and

HENRY F. DIAZ

NOAA/ERL/CDC, 325 Broadway, Boulder, CO 80303, U.S.A.

Abstract. It has frequently been suggested that the period encompassing the ninth to the fourteenth centuries A.D. experienced a climate warmer than that prevailing around the turn of the twentieth century. This epoch has become known as the *Medieval Warm Period*, since it coincides with the Middle Ages in Europe. In this review a number of lines of evidence are considered, (including climate-sensitive tree rings, documentary sources, and montane glaciers) in order to evaluate whether it is reasonable to conclude that climate in medieval times was, indeed, warmer than the climate of more recent times. Our review indicates that for some areas of the globe (for example, Scandinavia, China, the Sierra Nevada in California, the Canadian Rockies and Tasmania), temperatures, particularly in summer, appear to have been higher during some parts of this period than those that were to prevail until the most recent decades of the twentieth century. These warmer regional episodes were not strongly synchronous. Evidence from other regions (for example, the Southeast United States, southern Europe along the Mediterranean, and parts of South America) indicates that the climate during that time was little different to that of later times, or that warming, if it occurred, was recorded at a later time than has been assumed. Taken together, the available evidence does not support a *global* Medieval Warm Period, although more support for such a phenomenon could be drawn from high-elevation records than from low-elevation records.

The available data exhibit significant decadal to century scale variability throughout the last millennium. A comparison of 30-year averages for various climate indices places recent decades in a longer term perspective.

1. Introduction

"A thousand years ago, climate in the North Atlantic regions was perhaps 1 °C warmer than now..." (U.S. Dept. of Energy, 1989)

The spread of Norse seafarers and the establishment of colonies around the northern North Atlantic from Greenland to Newfoundland from about the close of the ninth Century A.D. has been taken as proof that the climate of this region was probably warmer than today. H. H. Lamb noted that in most areas of the world for which climatic conditions can be inferred, they seemed to have enjoyed, "a renewal of warmth, which at times [during the eleventh and twelfth centuries] may have

Climatic Change **26**: 109–142, 1994.

approached the level of the warmest millenium of post-glacial times." (Lamb, 1982, p. 162). The purpose of this special volume is to consider some of the available evidence for a warmer climate in different parts of the world during the time which has come to be known as the *Medieval Warm Period* (also referred to as the *Little Climatic Optimum* and *Medieval Warm Epoch*), namely from about the ninth to the fourteenth centuries A.D. Indeed, a major goal of our efforts has been directed to answering the question posed in the title of this review paper: was there a Medieval Warm Period (MWP), and if so, where and when?

As will be shown in the following sections, there is some evidence, including some recent studies based on tree-ring growth, that supports the occurrence of a warmer climate in the centuries preceding the start of the European Renaissance. The evidence comes primarily from three sources: growing season temperature reconstructed from tree rings (e.g. Cook *et al.*, 1991, 1992; Graumlich, 1993; Villalba *et al.*, 1990), inferred higher alpine glacier snow lines or glacier fronts (Grove and Switsur, this volume; Luckman, this volume), and from a variety of historical and phenological records which indicate the existence of a warmer climate compared to today's (Lamb, 1982; Zhang, this volume). Such periods of anomalously warm seasonal temperatures are only rarely synchronous between regions. A similarly diverse body of evidence indicates nothing exceptional taking place during this time in different parts of the world. For example, tree-ring evidence from the northern Urals (Graybill and Shiyatov, 1992) indicates only slightly higher summer temperatures in the first few centuries of this millennium. Evidence presented by Stahle and Cleaveland (this volume), also suggests that spring precipitation in the Southeast United States was not exceptional during either the ninth through fourteenth centuries, or during the 'Little Ice Age' (LIA) in the centuries that followed. Questions have been raised (for example, Alexandre, 1987; Ingram *et al.*, 1978; Ogilvie, 1991) regarding the reliability and representativeness of a significant portion of the historically-derived information which has been relied upon to infer climatic conditions for the North Atlantic-Western Europe region during the ninth through fourteenth centuries. Given these uncertainties, it seemed to us worthwhile to evaluate critically as much evidence as possible regarding the nature of climatic conditions during this period. One of the core projects of the International Geosphere Biosphere Program (IGBP, 1990) concerns the delineation of significant climatic and environmental changes that have occurred over the past 20,000 years, with particular emphasis on climatic variations during the last 2,000 years. The IGBP Core Project, referred to by the acronym PAGES, identified the so-called Medieval Warm Period as a focus for the detailed reconstruction of climatic parameters in comparison with present-day values. As an initial contribution to the process of evaluation and scholarly research focused on these goals, the authors convened a workshop to review critically the physical evidence for the existence of a significant climatic episode between the ninth to fourteenth centuries A.D., the time interval typically associated with the Medieval Warm Period. The meeting was held in Tucson, Arizona on November 5–8, 1991 with support from the National Oceanic and Atmospheric Administration (NOAA), the U.S. Department of Energy and the National Science Foundation. It provided an opportunity for interdisciplinary exchanges of information, and a forum to discuss and evaluate the scientific evidence from different sources and from different parts of the world regarding estimates of climatic differences in surface temperature and precipitation compared to twentieth century values. In the following sections, we give an overview of the physical, biological and historical evidence on the climate of the ninth through fourteenth centuries A.D.. The twelve original contributions, drawn from the Tucson workshop, that comprise this special volume of *Climatic Change* provide a diverse view of how different elements of the earth's climate have behaved in the last thousand years.

Clearly, improving our knowledge and understanding of past climate regimes, such as the 9th through fourteenth centuries A.D., may prove useful as potential predictors or analogues of future climatic patterns forced by increasing atmospheric greenhouse gas concentrations. We realize that a successful approach will require, *inter alia*, the participation of many experts in paleoclimatic reconstruction, and the generation of a variety of paleoenvironmental data. Together with modeling efforts to ascertain plausible sources of external or internal climate forcing mechanisms, such work will be crucial to assessing the validity and reliability of individual records and regional climate scenarios. It is our hope that this volume will serve as a catalyst to encourage further research on the reconstruction of the climate of this particular period, and to further our knowledge of natural climatic variations, whose understanding will be critical to the assessment of past, present, and future climatic changes.

2. The Medieval Warm Period

Several lines of evidence seem to suggest that much of northern Europe, the North Atlantic, southern Greenland and Iceland experienced a prolonged interval of warmth from about the tenth to thirteenth centuries A.D. exceeding conditions which prevailed during the early part of the twentieth century (Lamb, 1977, 1982). As this epoch belongs historically to Europe's so-called 'Middle Ages', the term, 'Medieval Warm Period' has come to be used to refer to the climate of this time. It may be noted that 'epoch', 'maximum', and 'Little Climatic Optimum' are terms also used to refer to this time interval. Nonetheless, contrary to the impression that the designation 'Medieval Warm Period' might give, there is much evidence of regional and seasonal inconsistency in climate in the ninth through fourteenth centuries. For example, Alexandre (1987) has carefully reviewed much of the evidence cited for inferring that northern Europe experienced a prolonged warm episode during that time. He concluded that the climate of southern Europe, i.e., the Mediterranean region, behaved quite differently from that of northern Europe, such that no exceptional winter warmth is evident in the available records for southern Europe until the mid-fourteenth century. On the other hand, Alexandre notes the predominance of warm springs and dry summers in western Europe throughout most of the thirteenth century (from A.D. 1220–1310). This is consistent with a phase of retreat of alpine glaciers around that time (see Grove and Switsur, this volume). The combination of cold springs and wet summers coupled with humid winters in the mid-fourteenth century could have favored glacial advances at that time.

Evidence presented in this volume indicates that the pattern of regional differences in the character of climatic anomalies during the ninth through fourteenth centuries, indeed shows spatial differentiation, in much the same way as the Little Ice Age (LIA), of the sixteenth to nineteenth centuries, does (Jones and Bradley, 1992). The papers by Luckman and Petersen (this volume) provide evidence to suggest that climatic conditions in the tenth to thirteenth centuries A.D. may have favored glacial retreat and forest advances to higher elevations in the Canadian Rockies, and possibly an improved crop growing environment in the U.S. Southwest around that time. Davis (this volume) reports evidence for greater lake depth or fresher water between A.D. 700 and 1350 in the region affected by the Arizona Monsoon, and points out that such an intensification of the monsoon could be consistent with higher solar receipts inferred from a radiocarbon minimum (Jirikovic and Damon, this volume). The study by Zhang (this volume) indicates a northward shift of phenological boundaries in China late in the period. On the other hand, the study by Stahle and Cleveland detailing a reconstruction of spring precipitation in the Southeast United States, fails to show any pronounced period (greater than 20 years) of either drought or wet spells in the ninth through fourteenth centuries. For that matter, the LIA also does not contain anomalous precipitation intervals longer than about two or three decades.

What kind of time scale might be characteristic of climatic events within a longer climatic epoch (spanning, say, a few centuries in length), such that it could be considered to differ, climatologically, from times immediately following and preceding it? To address this question we consider two types of evidence: the instrumental record of surface temperature in the Northern Hemisphere land areas (Jones *et al.*, 1986), and a suite of proxy temperature records derived primarily from tree-ring reconstruction of growing season temperature. We will discuss the development of the tree-ring temperature reconstructions in more detail in the next section.

Figures 1 and 2 show plots of the distribution of area-weighted winter (December–February) and summer (June–August) temperature anomalies for extratropical latitudes in the Northern Hemisphere (land areas poleward of 30° N) and tropics (equator to 25° N) for three time intervals – 1881–1920 (characteristic of the colder climate prevailing at the turn of the twentieth century), 1921–1979 (an intermediate milder period), and the recent warm decade of the 1980s. The graphs illustrate different spatial and temporal shifts in the median (and mean) value. For example, during summer extratropical latitudes warmed up the most from about 1900 to 1950, less so in the 1980s (Figure 1b). At higher latitudes in

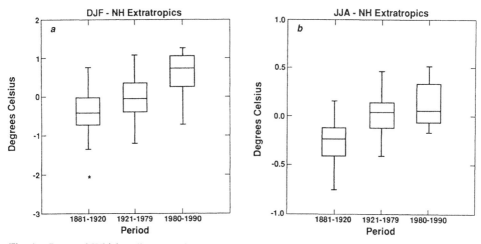

Fig. 1. Box and Whisker diagrams showing the distribution of land surface temperature anomalies (referenced to 1951–70) for the Northern Hemisphere extratropics (30 to 70° N) during three time intervals – 1881–1920, 1921–1979, and 1980–1990: (a) Winter (December–February); (b) Summer (June–August).

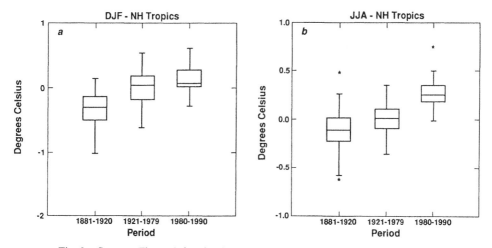

Fig. 2. Same as Figure 1, but for the Northern Hemisphere tropics (0 to 25° N).

winter (Figure 1a) and tropical latitudes during summer (Figure 2b) warming in the 1980s occurred at a much greater rate than during the first half of this century. Summer warming at the higher latitudes (Figure 1b) was much greater between the first and second period than during the 1980s.

There is no reason to believe that a similar heterogeneity of climatic signals (different rates of change according to season, latitude or region) was absent in the ninth through fourteenth centuries. Jones and Bradley (1992) make a similar point with regard to climatic variations since A.D. 1500. Diaz and Bradley (in press) also point out that the great variety of climatic signals and the inherent noise in the system (i.e., 'natural variability') makes it difficult to determine the presence of widespread climatic shifts on decade-to-century time scales. Below, we evaluate a range of the available records upon which characterizations of climate from the ninth to fourteenth centuries are based.

3. Tree-Ring Evidence

3.1. *Strengths and Limitations*

An *optimum* paleoenvironmental record of climate over the last thousand years or more would have the following six characteristics:

1. it would be continuous through the whole millennium and preferably longer, with dating good to the calendar year;
2. well understood spatial applicability;
3. a strong and well-defined climate signal of known seasonal applicability;
4. a temporal resolution of one year or better;
5. time-invariance in strength and nature of climate signals; and
6. be capable of recording century-scale as well as interannual and decadal scale fluctuations.

In addition, the optimum type of record would not be unique to one location, but rather would have a sufficiently widespread geographical distribution to capture large scale fluctuations consistently over a large region. This is of particular importance given the possibility that spatial representativeness may be frequency-dependent.

The primary focus here will be on statistically calibrated and verified reconstructions of climate from tree rings. There are a number of published tree-ring based reconstructions of climate in the last millennium with all or most of these characteristics. While the available reconstructions are drawn from a wide range of locations, none are at low latitude and there are far too few for them to be representative of any large region of the Earth. Given the information presented in Figures 1 and 2, this last observation has important implications regarding generalizations of individual (regional) climate signals to much larger scales.

Earlier examinations of the climate of the last millennium used tree-ring chronologies from which climate fluctuations were inferred, some after careful consideration of the processes limiting tree growth (e.g., LaMarche, 1974) and several apparently without any such analysis (e.g., Lamb, 1977; Ladurie, 1971; Bryson and Julian, 1962). Given the recent growth in availability of statistically calibrated and verified reconstructions in the literature there is neither merit in the use of uncalibrated time series, nor is there any need to continue this practice except in very limited circumstances. Thousand year or longer calibrated and verified reconstructions of seasonal or annual temperature are available for the Polar Urals (POL) (Graybill and Shiyatov, 1989), northern Fennoscandia (FEN) (Briffa *et al.*, 1992), northern Italy (Serre-Bachet, this volume), gridpoint reconstructions in western

Europe based on a combination of tree-rings and other data (EUR) (Guiot *et al.*, 1988; Guiot, 1992), the Sierra Nevada, California (SNW) (Graybill, 1993); (SNS) (Graumlich, 1993); (Scuderi, 1993), Tasmania (TAS) (Cook *et al.*, 1991, 1992), and southern South America (RIA) (Villalba, 1990, and this volume; Lara and Villalba, 1993). Similarly derived reconstructions of precipitation or precipitation related variables such as drought indices have been reported for the Sierra Nevada, California (Graybill, 1993), the Colorado Plateau (Rose *et al.*, 1981; D'Arrigo and Jacoby, 1991; Taylor *et al.*, 1992), the southeastern United States (Stahle *et al.*, 1985; Stahle and Cleaveland, 1992; Stahle and Cleaveland, this volume), Morocco (Till and Guiot, 1990; Chbouki, 1992), and Chile (Boninsegna, 1990). In addition, a 2100 year record of extreme droughts in the San Joaquin Valley in California has been derived from giant sequoia tree rings (Hughes *et al.*, 1990; Hughes and Brown, 1992).

The authors of the temperature reconstructions mentioned above have, to a large extent, used the same or at least comparable methods in their derivation. Similarly, they all address explicitly all or most of the characteristics of an ideal reconstruction listed at the beginning of this section. All these reconstructions, except EUR, extend through the whole of the millennium, are dated to the calendar year and have a temporal resolution in their climate signal of a year or better. Efforts have been made to address the strength, nature and seasonal applicability of their climate signals. In some cases, there is an explicit discussion of the spatial applicability of the climate signal (e.g., Graybill, 1993), and in all cases they document at least the locations of the trees sampled. Special attention will be given here to the following issues:

1. the strength of the climate signal and its geographic applicability for the reconstructions available to us;
2. the time-invariance of the strength and nature of the climate signal;
3. the extent to which these reconstructions faithfully record century-scale as well as interannual and decadal fluctuations.

Table I gives the linear correlation between the tree-ring based reconstructions selected for study and the nearest ensemble of available long-term instrumental records. It is clear from this table that there is considerable variation in the extent to which each of the six entirely tree-ring based temperature reconstructions, (filtered versions of which are shown in Figure 3), is correlated with regional instrumental temperature records. There is a much greater number of stations available within 500 km of the locations for the Sierra Nevada reconstructions (SNW and SNS) than for the other reconstructions. Thus, the influence of the individual station or stations used to calibrate each record is much diluted in their particular case, as compared to the others. Taking into account the calculations and careful assessments of the climate signal and its reliability by the original authors of the records, along with the results shown in Table I, it may be concluded that these reconstructions contain considerable information about the temperature variations in their respective region at radii of several hundred kilometers. The proportion of the variance of

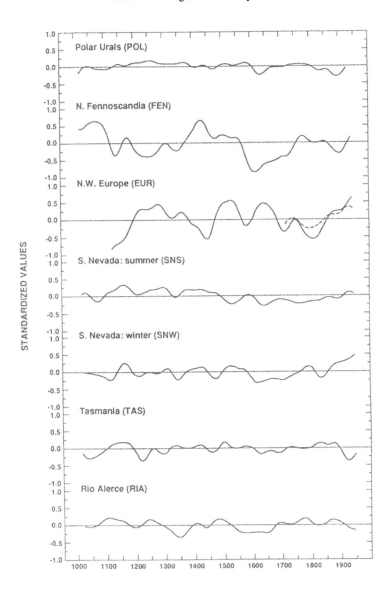

Fig. 3.   Low-pass filtered time series for seven long-term temperature reconstructions. Filter has a 50% frequency cut-off at 100 years and is used to highlight century time scale variations in the data. See text for a description of data and sources. The amount of variance present in each series on this time scale is as follows: (a) Polar Urals (POL) – 2.5%; (b) northern Fennoscandia (FEN) – 21.6%; (c) N.W. Europe (EUR) – 26.1%; (d) Sierra Nevada summer (SNS) – 4.0%; (e) Sierra Nevada winter (SNW) – 6.0%; (f) Tasmania (TAS) – 4.4%; and (f) Rio Alerce (RIA) – 3.3%. EUR is the mean of the time-series reconstructed for the four grid points at 50 and 55° N and 0 and 10° W, being the strongest grouping of high calibration statistics in the region reported by Guiot (1992). Dashed line plotted with EUR curve illustrates low-pass filtered Central England winter temperature in the period of overlap. It accounts for 6.7% of the total Central England winter temperature variance.

TABLE I: Correlation between tree-ring based temperature* reconstructions and instrumental records interpolated (using inverse distance weighted method) to the location of the tree-ring reconstruction using search radii of 500, 300 and 200 and 100 km

Reconstruction	Correlation	Number of years	Number of stations
POL Polar Urals	**0.75/0.75/0.76/0.76**	87/87/87/87	2/2/1/1
FEN Fennoscandia	**0.77/0.79/0.71/0.76**	111/111/72/30	7/4/2/1
SNS Sierra summer	0.32/**0.44/0.41/0.44**	119/119/119/110	33/21/13/6
SNW Sierra winter	**0.38**/*0.25/0.21/0.25*	118/118/118/110	33/21/13/6
TAS Tasmania	**0.40/0.44/0.42**/*0.59*	120/114/114/26	5/4/4/1
RIA Rio Alerce	**0.54/0.57/0.58/0.56**	53/53/53/53	6/4/3/1

* seasonal mean values of station temperatures corresponding to the season of the tree-ring reconstruction were used.
Bold signifies $p < 0.0001$.
Plain signifies $0.0001 < p < 0.001$.
Italic signifies $0.001 < p < 0.05$.

regional-scale temperatures captured by the reconstructions ranges from over 55% for the Polar Urals (POL) and northern Fennoscandia (FEN), through 30 to 42% for Rio Alerce (RIA) and 20 to 38% for Sierra Nevada summer temperature (SNS), down to 16 to 35% for Tasmania (TAS) and 10 to 38% for Sierra Nevada winter temperatures (SNW). In some cases the original authors of these reconstructions report stronger correlations with instrumental data. It should be remembered that they will normally have chosen the meteorological station or stations they judged to be most appropriate for comparison with their tree-ring based material, for example, in terms of the elevation of the stations. We have made no such selection in the calculations reported in Table I. The 1000 years or longer temperature reconstructions mentioned above, but not shown in Table I, appear to be broadly similar in terms of the strength of their climate signal. Serre-Bachet's (this volume) reconstruction of summer temperatures in northern Italy is reported as capturing at least 33% of observed temperature variance in the period of overlap with instrumental records; Guiot's European reconstructions (EUR) based on tree-rings and other kinds of records capture a mean of 40% of temperature variance; the southern South American reconstruction (RIA) captures 33 to 37%; and Graybill reports a range of explained variance in verification periods of from 34 to 64% in his Sierra Nevada temperature (SNW) reconstruction. Most of these reconstructions are based on relatively limited amounts of material from one to a very few sites, and consequently have weaker climate signals than those obtained from multi-site tree-ring networks for which material of only 500 years' length or less is available. Such shorter reconstructions, especially when wood density rather than ring width is the main variable used, often account for as much or more temperature variance than the 55% plus seen here in the Polar Urals (POL) and in Fennoscandia (FEN) (Briffa *et al.*, 1988; Briffa *et al.*, 1992; Conkey, 1986; Hughes and Davies, 1987; Hughes, 1992). It is very likely that the quality (for example, the proportion of temperature variance accounted for) of most existing 1000-year temperature reconstructions other than the two northernmost (POL and FEN) would be improved if further multiple site chronologies were available.

There are four major sources of change in the strength and nature of a given climate signal in a reconstruction. First, the number of radii series and trees contributing to each chronology changes through time, usually being smallest at the beginning of the chronology. Most authors take account of this in deciding from which beginning year to report their reconstruction. This is not a significant problem with most of the reconstructions discussed here, at least for the last 1000 years. Similarly, there may be a change in the number of chronologies used in deriving the reconstruction. In fact, some of the reconstructions discussed here are based on single site chronologies (e.g., RIA, TAS, SNW). Others are based on multiple-site mean chronologies for a small region (FEN and POL), while yet others are based on groups of separate site chronologies (SNS, EUR). Reconstructions based on regional mean chronologies, or on groups of separate site chronologies, are much less likely to be distorted by non-climatc changes than those based on a single site chronology, since changes unique to a single site are to some degree diluted when material from multiple locations is used. Reconstructions based on groups of separate site chronologies have the further advantage that they have the potential to capture information related to larger scale climate phenomena. This is for two reasons: if the chronologies are sufficiently widely distributed, they will respond to the various sub-regional effects of particular climate fluctuations and events, and, as already mentioned, site-specific effects, including microclimate, may be diluted in a multi-chronology set.

The second source of change in climate signal strength and nature is associated with tree biology and ecology. The first few decades of a tree's life are marked by physiological and developmental differences from later life. As they only apply to a very short period (at most a few decades) these changes are likely to be relatively unimportant for most of the reconstructions discussed here, excepting for the first few decades of those that actually begin close to A.D. 1000. Other changes in tree-ring response to climate are associated with the trees' relative position in the forest canopy, which usually stabilizes once each tree has completed most of its height growth. This is unlikely to be important in the case of chronologies built from trees growing in open stands, i.e. those where the trees are so far apart that there is little or no interaction between them (e.g., POL, FEN, all the Sierra Nevada temperature reconstructions and the Graumlich Sierra Nevada precipitation reconstruction). For chronologies from more closed forest stands where greater inter-tree interaction might be expected, this problem is to some degree dealt with by the process of standardization (see below).

The third source of change in the climate signal concerns site conditions. These changes might arise from gradual changes in soil chemistry or in site hydrology, or from more acute changes associated with changes in stand structure caused by disturbance such as fire or human intervention. The influence of disturbance on site chronologies can be minimized by careful site and tree selection, and by inspection of cores for evidence of disturbance and their exclusion from the site chronology, e.g. sudden growth surges or suppression, reaction wood, traumatic resin ducts, fire scars, and so on. This is standard practice in dendroclimatology. It is more difficult to guard against the influence of gradual site changes on climate reconstructions, except again by the use of material from multiple locations. A particular problem arises when trees from an area close to a distributional limit such as elevational or latitudinal treeline are used. Graybill and Shiyatov (1992) discuss this issue with respect to their Polar Urals (POL) temperature reconstruction, and conclude that the evidence of tree line movement contained within their data is consistent with the general pattern of their reconstruction.

The fourth source of noise in the recorded climate signal arises from the interaction of the various climate factors that might control tree growth. For example, the influence of summer temperature on ring width in semi-arid high elevation sites may only be evident in unusually wet years. A shift in the frequency of such years would produce variations in the relative influence of these two factors on ring width. It is conceivable that such changes could cause the instrumental period in which the reconstruction was calibrated and verified to have few analogues in the last 1000 years. This possibility has been dealt with in a number of ways. One is to choose regions, sites and tree ring variables that have a simple and dominant relationship to a climate variable of interest. Examples of such combinations are conifer maximum latewood density from regions with cool moist summers as a summer temperature recorder, and ring widths from mid- to low elevations in semi-arid conditions as recorders of soil moisture or seasonal precipitation. Another approach is to deal explicitly with these interactions, which may be non-linear, for example by developing response surfaces (Graumlich, 1993). Finally, the ensemble of chronologies with different climate responses within a region can be used (Fritts, 1991) as a transducer capable of capturing a wider range of interactions than any 'single response function' data set.

How well do these temperature reconstructions represent century scale fluctuations? There is ample evidence from the statistical calibrations and verifications discussed above that the reconstructions contain significant amounts of information on relatively high-frequency fluctuations (two or three decades down to interannual, see Diaz and Pulwarty, this volume). Long-term tree-ring changes associated with tree age and stand dynamics have been removed from the chronologies used in these reconstructions by a variety of techniques. The problem is to remove these non-climatic changes from the tree ring series while retaining as much climate-induced variability as possible. In the case of almost all the temperature reconstructions discussed here, strenuous efforts have been made to conserve climate-related variation on time scales of up to one or a few centuries. Generally, this has involved using core or tree series several centuries long and detrending very conservatively using a straight line, negative exponential or very stiff spline. In the case of the Fennoscandia (FEN) reconstruction, a rather simple standardization curve was derived from the date-independent mean growth curve of a large sample of trees (Briffa *et al.*, 1992). It does seem likely, however, that the standardization method used by Serre-Bachet (this volume), namely, the ratio of measured ring-width to a curve produced by a 60-year low-pass filtering of the data, would minimize all century-scale features in the chronologies, including those due to climate. Guiot *et al.* (1988, 1992) has used similar methods to Serre-Bachet but his reconstructions contain century-scale information from the ice core and historical data he uses along with tree rings. Finally, the possibility should be recognized that some trees may be able to adapt to century-scale climate fluctuations (see Stahle and Cleaveland, this volume) and hence will not record such changes in their tree-ring series.

3.2. *Interpretation of the Climate Record*

What evidence of a Medieval Warm Epoch, if any, is offered by these reconstructions? Figure 3 illustrates the time behavior of the temperature reconstructions considered here. The annual data have been standardized for ease of comparison and smoothed with a Lanczos filter (Duchon, 1979) having a frequency cutoff of 50% at 100 years to emphasize variation with periods of 100 years or longer. It is of interest that the variance passed by these filters is, in most cases, only a small fraction of the total variance of the unfiltered series, usually less than 5%. This is similar to the fraction of total variance passed by an identical filter applied to the longest available instrumental temperature record, Manley's Central England series. Winter temperatures for the period from 1721–1990 were used in this calculation. Two reconstructions show proportionately much more century-scale variance, namely summer temperatures at Torneträsk, northern Sweden (FEN), and annual temperatures in northwestern Europe (EUR). It is not clear whether this higher proportion of century-scale variance in these two reconstructions represents local climate reality or is the product of the methods used to develop them. Summer temperatures in the Polar Urals (POL) and the Sierra Nevada (SNS) show persistent positive anomalies from the millennium mean from around A.D. 1110 to 1350 and A.D. 1090 to 1450, respectively. There is a persistent warm anomaly in Guiot *et al.*'s (1988, 1992) multiproxy reconstruction of northwest European annual temperatures between A.D. 1190 and 1350, whereas the Fennoscandian reconstruction (FEN) shows negative anomalies at the same times as the positive anomalies in the Polar Urals (POL). The largest persistent positive anomalies in the Fennoscandian reconstruction were from A.D. 971 to 1100 and from A.D. 1350 to 1540. A positive anomaly centered around A.D. 1150 is seen in both Sierra Nevada reconstructions in Figure 3 (SNS and SNW) and in Tasmanian summer temperatures (TAS). A similar anomaly is seen in southern South American summer temperatures (RIA) centered around A.D. 1090, but overlapping the warm anomalies in SNS, SNW, and TAS. These are the largest warm anomalies in SNS, TAS and RIA, and the second largest in SNW. For all series shown in Figure 3 the warm anomalies between A.D. 900 and 1300 have only been exceeded subsequently in two cases, the twentieth century in Sierra Nevada winter temperatures (SNW) and EUR. There is, however, only limited synchrony in the anomalies in the tenth to fourteenth centuries. Serre-Bachet (this volume) notes as many cold as warm episodes between A.D. 1000 and 1300, with no evidence of warming in the twelfth century (although the probability that her reconstruction may not reflect century scale variations adequately was discussed above). It has already been noted that Guiot *et al.*'s (1988, 1992) multiproxy reconstruction of northwestern European temperatures (EUR) shows temperatures close to modern between A.D. 1200 and 1300, colder before 1200 and between 1370 and 1450, and similar to present in the early sixteenth and early seventeenth centuries. Lara and Villalba (1993) do not find evidence for a Medieval Warm Period in their long summer temperature reconstruction from southern South America, although they may have reduced their chances of observing such a change by standardizing their tree-ring series using a spline with a 50% variance reduction at 128 years. Scuderi's (1993) reconstruction of annual temperatures in the Sierra Nevada indicates a warmer period from the late tenth to early twelfth centuries, but this is not as marked as a warm period in the sixteenth century. There is evidence to support the low frequency trends in Graumlich's (1993) reconstruction of summer temperatures in the Sierra Nevada (SNS) by virtue of its broad similarity to the growing season temperature changes inferred by LaMarche (1974) from upper elevation bristlecone pine ring widths from the White Mountains, 50 km or more to the east of the Sierra Nevada sites. LaMarche's speculation that this series might represent temperatures, at least on multidecadal and longer time scales prior to the mid-nineteenth century, has received support recently from the similarity of a series derived from upper elevation bristlecone pine ring widths in a larger region, and a smoothed version of a well calibrated summer temperature reconstruction derived from wood density from a completely independent network of conifer chronologies (Graybill and Idso, 1993). The similarity between Graumlich's reconstruction (SNS) and LaMarche's series suggests that the pattern of a sustained Medieval Warm Epoch followed by a Little Ice Age illustrated in a number of technical and popular works on past climate may well have existed at high elevations in the mountains of eastern California. Further support is given by the existence of the same pattern, particularly the cool period in the seventeenth and early eighteenth centuries, in Graybill's Sierra Nevada winter temperature reconstruction (SNW). It is of interest that all the tree-ring materials used by these authors (Graumlich, Graybill and LaMarche) in the White Mountains and Sierra Nevada were collected at high elevation (3000 m, or above).

It is instructive to view these 1000-year reconstructions in terms of consecutive 30-year periods commonly used in descriptive climatology. Figure 4 (a through

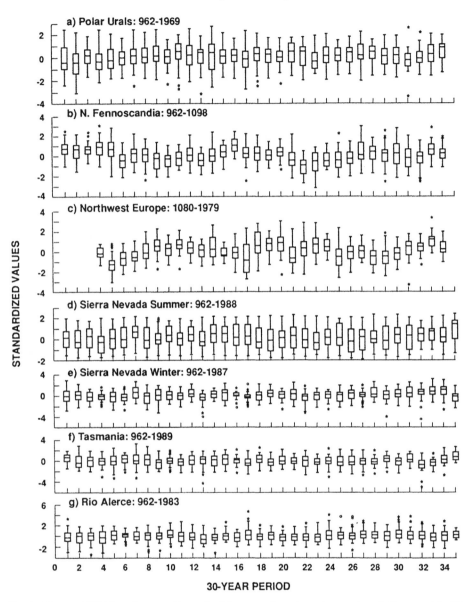

Fig. 4. Box and Whisker diagrams for each of the seven long-term temperature indices used in this study. Graphs show the distribution of standardized values (over the record length of each series) in each successive 30-yr interval. See text for a description of data and sources.

g) shows the seven series featured in Figure 3 in terms of box-and-whisker diagrams for non-overlapping 30-year periods commencing in A.D. 962. Each box gives the distribution of the central 50% of the standardized values (i.e., values in the range between the 25th and 75th percentiles); the horizontal line in each box represents the median, the vertical lines (whiskers) extend out from

TABLE II: Calendar year intervals for the 35 thirty-year periods shown in Figure 4 Period covered is A.D. 960 to 1989. Note that the last interval, number 35, covers only 8 years 1982–1989

Index no.	Calendar years	Index no.	Calendar years	Index no.	Calendar years
1	962–991	8	1172–1201	15	1382–1411
2	992–1021	9	1202–1231	16	1412–1441
3	1022–1051	10	1232–1261	17	1442–1471
4	1052–1081	11	1262–1291	18	1472–1501
5	1082–1111	12	1292–1321	19	1502–1531
6	1112–1141	13	1322–1351	20	1532–1561
7	1142–1171	14	1352–1381	21	1562–1591
22	1592–1621	29	1802–1831		
23	1622–1651	30	1832–1861		
24	1652–1681	31	1862–1891		
25	1682–1711	32	1892–1921		
26	1712–1741	33	1922–1951		
27	1742–1771	34	1952–1981		
28	1772–1801	35	1982–1989		

the 25th and 75th percentiles up to 1.5 times the interquartile range (IQR) and the starred and circled values correspond to annual values at distances between 1.5 and 3.0 times the IQR, and greater than 3.0 times the IQR, respectively. The highest median in POL is for 1952–1969, the last 18 years in the record, followed closely by 1262–1291. Table II identifies the fixed 30-year period (starting in A.D. 962) corresponding to each index value plotted along the abscissa of each graph. The last '30-year box' may contain less than 30 values depending on when the particular series ended. Beginnings of record boxes may also contain fewer than 30 values. In FEN, the highest median is for 1412–1441, followed by 1052–1081, and then by a group of 30-year periods in the tenth and eleventh centuries. The warmest median for EUR is from 1922–1951 followed by 1532–1561, 1502–1531 and 1262–1281. The highest 30-year median in SNS is for 1142–1171 followed by 1352–1381 and 1382–1411, then 1922–1951. In SNW, the highest median is for 1952–1981, followed by 1142–1171 and 1382–1411. The periods 962–991 and 1112–1141 have the highest 30-year medians in Tasmania (TAS), followed by a group of periods from 1322 to 1411. At Rio Alerce (RIA) 1112–1141 has the highest 30-year median followed by 1082–1111 and several other periods in the first half of the millennium. In only two out the seven cases illustrated in Figure 4 does the first half of the seventeenth century not contain one or more of the four coldest 30-year periods in the millennium, and three have their coldest 30-year period in this half-century. In no case where a full 30 years data are available for 1952–1981 does this period stand out as unusual when compared with the rest of the millennium (see Diaz and Bradley, in press).

Is there any distinguishing feature of the Medieval period to be found in any of the tree-ring reconstructions of precipitation? Graybill (1993) reconstructs multi-decade periods of low winter precipitation in the Sierra Nevada around 1300, 1600 and 1900, followed on each occasion by wetter than normal conditions. There is some similarity between this reconstruction and that produced independently by Graumlich (1993) for the same region, who notes drought periods between A.D. 800–859, 1020–1070, 1197–1217, 1249–1365, 1443–1479, 1566–1602, 1764–1794, 1806–1861 and 1910–1934. She finds no evidence of century-scale or longer deviations in her reconstruction. Reconstructed periods of higher frequency of extreme drought events in the San Joaquin drainage (Hughes and Brown, 1992) are inferred from A.D. 700 to 850, 250 to 350 and 1480 to 1580 in order of decreasing intensity, and of relatively low frequency droughts from 100 B.C. to A.D. 100, 400 to 500, 1600 to 1700 and 1850 to 1950. If distinguished at all, the Middle Ages demonstrate a relative lack of multi-decade or century scale variation in the frequency of these extremes. The only significant exception to this is the unusually wet period between A.D. 1080 and 1129 inferred from anomalously low $\Delta^{13}C$ in bristlecone pine in the White Mountains of California (Leavitt, this volume). The Medieval Period does not appear to differ from the rest of the last millennium in reconstructions of precipitation on the Colorado Plateau (Taylor *et al.*, 1992), except that the last authors note abrupt changes in the power spectrum between A.D. 1280–1350, 1416–1534, 1645–1714 and 1800–1835, all believed to be periods of low sunspot activity. In particular, they report that oscillations with periods of approximately 20 and 80 years are absent during times of low sunspot activity. Stahle and Cleaveland (this volume) observe no century scale effects in reconstructed spring precipitation in the southeastern United States, although the second to fourth driest decades did occur between A.D. 900 and 1300 (945–954, 1192–1201, 1082–1091, 1262–1271; the driest was 1756–1765). They do note a relative prevalence of decadal scale dry episodes between the mid-eleventh and early thirteenth century. Similarly, Moroccan winter precipitation (Till and Guiot, 1990) tended to be low between 1186 and 1234, and from 1379 and 1428, whereas wet years were unusually frequent between 1250 and 1300. A sustained drought is reconstructed for the period 1270–1450 at Santiago de Chile (Boninsegna, 1992), followed by another such episode between 1600–1650. This reconstruction is considered highly reliable after 1250. It is difficult to discern any consistent multi-decade or century-scale pattern in the available precipitation reconstructions.

3.3. *Summary of Dendroclimatology Evidence*

Other than the regions containing the Sierra Nevada and the Great Basin, northern Fennoscandia, the Polar Urals (POL) and Guiot's large regional reconstructions (whose low frequency component may be substantially affected by less well dated and more poorly understood records than tree rings), these reconstructions offer little support for a *sustained* Medieval Warm Epoch/Little Ice Age sequence in the last 1000 years. On the other hand, there are some consistent features such as: (a) an increased tendency toward cold conditions in the early and mid-seventeenth century; and (b) either a warm period in the mid-twelfth century and/or the late thirteenth/early fourteenth centuries. This latter feature is not striking. The records for the twentieth century show no consistent distinguishing features in comparison with the rest of the millennium (with the exception of the Sierra Nevada winter temperature reconstruction). It should be noted that only TAS covers the whole of the 1980s, while POL and EUR do not cover this decade at all, and FEN ends in 1980.

4. Ice Core Evidence

4.1. *Strengths and Limitations*

Annually layered ice cores from the polar regions or from high elevations at mid- and low-latitudes may possess a number of the characteristics of an optimum paleoclimate record of the last thousand years listed above (section 3a). There are many situations in which records spanning the last millennium, and indeed much longer, have apparently annual layers. In general, the occurrence of such records depends upon a sufficiently reliable and high rate of deposition to allow identification of seasonal features, and on maximum temperatures being too low to permit melting. Almost all the published records are based on single cores, relying on the distinctness of interannual features such as seasonal cycles of stable isotopes (usually ^{18}O), conductivity, and concentrations of ions and of microparticles as guides to the identification of annual layers. Having identified the annual layers, the investigators establish a chronology by counting them back from the known date of sampling. This procedure fails to deal with situations in which one or more annual layers has been lost, perhaps by wind action, or where a strong intra-annual feature has been mistakenly identified as an annual layer. Thompson (1991) points out that "It is important that two ice cores be drilled at the same site and independently analyzed to provide an independent verification of the time scale and to ensure an uninterrupted physical and chemical stratigraphic record" (p. 205). Of course, it is possible that a dating error could survive even this approach if the same layer was missing from both cores. In ideal circumstances multiple cores would be used. It is sometimes possible to use a specific marker event to check an ice-core chronology. For example, a notable dust layer in the Summit Core from Quelccaya, Peru has been identified with the historically recorded eruption of Huaynaputina from 19 February to 6 March A.D. 1600 (Thompson, 1991). In cases where such markers are not available, and in particular in the earlier part of the millennium in records based on a single core, dating errors of several years or more cannot be ruled out. This limits the value of such records in the present investigation, where decadal-scale variation is important.

Few direct analyses of spatial relations between interannual variability in ice-core measurements, usually $\delta^{18}O$, and meteorological records have been made that would allow a rigorous assessment of the spatial applicability of these records. Lack of consistency has been noted between stable isotope time series from different parts of, for example, Greenland (Robin, 1981) or the Antarctic Peninsula (Jones et al., 1993). Some of this inconsistency may be ascribed to intraregional climate complexity, and some to noise in the ice core records. The paucity of meteorological stations in polar and high-elevation regions makes it impossible to do the kind of analysis presented for tree ring-derived temperature records in Table I. The best hope for gaining a firmer understanding of the spatial applicability of ice core derived records of, for example, temperature and snowfall, may rest in a combination of regional networks of cores and partially synthetic meteorological data based on forecast models with adequate mesoscale resolution.

Annually layered ice core records are, of course, absent from the greater part of the Earth's surface. They are found, however, in several regions of great importance to understanding of the global climate system, where few if any other records exist, notably the polar ice caps, the Tibetan Plateau, and the High Andes. Of the many variables that may be measured in the annual layers of ice cores, $\delta^{18}O$ has received the greatest attention as a potential source of information on past temperatures. This variable is often assumed to be a recorder of "changes in the temperature gradient from the vapor source region(s) to the ice core site" (p. 96 in Oeschger and Langway, 1989), or of temperature at the ice core site. As Jones et al. (1993) p. 14 point out, "Although the isotopic fractions of the heavier oxygen-18 and deuterium in snowfall are temperature dependent, researchers should not unquestioningly add a temperature scale to isotope time-series plots. The temptation to do this arises from the strong spatial correlation between the annual-mean temperature and the mean isotopic ratio ($\delta^{18}O$ or δD) of precipitation. This spatial correlation, however, masks the complexities of the isotopic evolution process, from the source through atmospheric transport to final deposition. . . Because of these many problems one cannot assume a priori that isotopic changes at a single site are closely related to temperature changes." They go on to recommend that isotope based reconstructions be based on relationships that have been checked against instrumental climate data on appropriate time scales. This has not been done very often, due no doubt in part to the lack of long-running meteorological stations near the sites of ice cores. In the few cases where such analyses have been attempted (for example Jones et al., 1993, Robin, 1981) only about one-third of annual temperature variance on time scales of years to decades is accounted for by the isotopic data. In several cases this estimate is based on 30 or fewer years and hence is of little value in assessing these records on the decadal time scale. The alternative approach of comparing isotopic records with temperature histories derived from borehole or ice temperatures has been proposed (Robin, 1981), and is to be applied on time scales relevant to this review as part of the GISP2 project (Alley and Koci, 1990).

Although it is clear that the seasonal nature of isotopic signals in ice depends in part on the seasonal distribution of snowfall, which may be very short, it could also be affected by seasonal variations elsewhere between the vapour-source region and the site of the ice core. Changes in ocean currents, seasonality of snowfall and storm-tracks, for example, could lead to changes in the isotope record over decades and centuries. The use of deuterium excess above that calculated from a temperature-determined relationship between $\delta^{18}O$ and δD to detect changes in source region is possible (Oeschger and Arquit, 1991). This might be a useful component of a screening procedure for time-variance in the nature of the climate signal in $\delta^{18}O$ time series.

It is difficult to assess if isotopic records from annually-layered ice cores are capable of recording century-scale as well as interannual and decadal fluctuations, given the uncertainties surrounding the nature and strength of the climate signal. There are, however, some cases where marked shifts in the relationships between a suite of variables measured in an ice-core occur almost simultaneously and persist for several decades or longer (Thompson and Mosley-Thompson, 1989; Mayewski *et al.*, 1993). This might represent a more general change in atmospheric conditions than should be deduced from the variation of a single quantity.

4.2. *Evidence for a Medieval Warm Epoch*

Inconsistencies between $\delta^{18}O$ time series from several locations in the region including Greenland were noted by Robin (1981, p. 193), who pointed out that "a warm spell ... in the middle of the twentieth century is almost the only feature common to all records", inferring warmth from less negative $\delta^{18}O$ values. Decadal scale differences between regions within Greenland are known from the instrumental record, and so it is not particularly surprising that such differences are seen in the isotopic time series. On the other hand, Ogilvie (in press) has shown similarities between multi-decade anomalies in the Crête and Milcent $\delta^{18}O$ series, Dye-3 melt data, Iceland sea ice data and Iceland temperatures reconstructed from historical documents for the period A.D. 1600 to 1800.

There is little evidence of a globally coherent pattern of century scale variation in the north-south global transect assembled by Thompson (1991) and reproduced in Figure 5. The time series of decadal averages of $\delta^{18}O$ have been interpreted as temperature records. Only two, those from Quelccaya in the Peruvian Andes and from South Pole, Antarctica, show evidence of a higher incidence of above average 'temperatures' in the first part of the 1000 years followed by a period of more frequent below-average decades after A.D. 1520. Quelccaya Summit is at 5670 m and South Pole is at 2912 m, whereas Siple Station is at 1054 m. The other high elevation core included in Figure 5, Dunde from the northern Tibetan Plateau, does not show this pattern. Dansgaard *et al.* (1975) noted a strong similarity between smoothed versions of the Crête $\delta^{18}O$ series and Lamb's English temperature record based on a mixture of historical and other sources, and early instrumental records. There is considerable similarity in these smoothed series during their most reliable

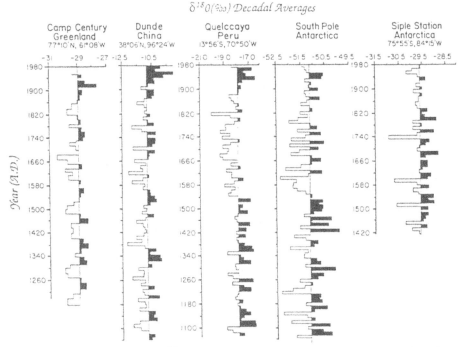

Fig. 5.   Decadal averages of the δ^{18}O records in a north-south global transect. The shaded areas represent periods isotopically less negative than the long-term mean for that core as shown here, which is commonly interpreted as indicating relative warmth. The converse applies to the unshaded areas. Reproduced from Thompson (1991).

period, that since A.D. 1700. The Crête series does indicate a higher frequency of above average 'temperature' multidecadal periods in the ninth and eleventh centuries than at any time since, except the twentieth century. It may be relevant to note that Crête is the highest elevation Greenland ice core (3172 m) discussed so far in this review. For comparison, Camp Century is at 1885 m.

4.3. Summary of Ice Core Evidence

Considerable caution should be exercised when interpreting time series of δ^{18}O as temperatures on decadal to century time scales, although multi-variable analyses show promise. There is evidence for higher frequency of above-average δ^{18}O on decadal and multidecadal time scales at the end of the first millennium A.D. and in the early centuries of the second millennium A.D. at Quelccaya, Peru, South Pole and Crête, Greenland, all sites at close to 3000 m elevation or higher.

5. Documentary Evidence

5.1. *Strengths and Limitations*

H. H. Lamb (1977, p. 435) drew the following conclusions concerning the 'medieval warm epoch, or Little Optimum'

> "In the heartland of North America, as in European Russia and Greenland, the warmest times may be placed between about A.D. 950 and 1200. In most of Europe the warmest period seems to have been between 1150 and about 1300, though with notable warmth also in the late 900s. In New Zealand the peak may have been as late as 1200 to 1400. In southernmost South America the forest was receding rapidly to western aspects only, indicating more effective rain shadow from the Andes."

Lamb was able to collate a much higher geographic and temporal density of evidence on the climate of Europe, particularly northern Europe, than for any other region. This applies especially to documentary sources. He brought together a very wide range of evidence for the existence, location and nature of this 'warm epoch', including natural archives such as tree rings, ice cores, and speleothems, paleoecological analyses (notably those concerning latitudinal and longitudinal distribution of forest vegetation, and of inshore marine conditions), geomorphological analyses (particularly concerning glaciers and fluvial processes), archeological evidence (particularly as it refers to the latitudinal and elevational extent of cultivation of various crops), written records of various kinds referring to meteorological phenomena (especially extreme seasons, but also such matters as observed sea ice), sea level changes, phenological records, crop prices and social conditions.

These records were used to establish a time course of relative change for a number of features of seasonal climate at particular locations or in specified regions, for example, an index of summer dryness/wetness in England based on documentary sources. Various methods were then used to quantify, in terms of meteorological measurements, the changes inferred from the analysis of this very large collection of information. A notable example of this is an assessment of crop or other plant distributional data and wine harvest dates, all from western Europe. A table was presented (Lamb, *op. cit.*, Table 17.1, p. 426) of calculated temperature departures for the warmest month of the year for various parts of the Middle Ages compared with 'average values prevailing about 1900'. These ranged from 0.5 to 1.6 °C and seemed to refer primarily to temperatures in some period of unspecified length ending around the early fourteenth century, although the change in wine harvest dates is reported as taking place after 1550. Other than this table, the major evidence Lamb (*op. cit.*) presented for the level of temperature change associated with the 'medieval warm epoch' is based on reports or inferences concerning conditions in the Viking colonies in Greenland.

Knowledge of many of the types of records used by Lamb in his earlier papers (Lamb, 1965) and in his classic 1977 monograph, as well as by others focussing on this period (e.g., Flohn, 1950) has advanced by a considerable extent in the last decade. In the case of evidence based on documentary sources, the most important methodological development has been the systematic and rigorous application of the techniques of historical analysis as advocated, for example, by Gottschalk (1971–1977), Ingram *et al.* (1978), Bell and Ogilvie (1978), Ingram *et al.* (1981), and Pfister (1988). These techniques emphasize a dependence on contemporaneous reports of well established provenance and a critical review of the reliability of the record based on the highest standards of modern historical scholarship. Such analyses (e.g., Alexandre, 1987; Ogilvie, 1991) have led to the rejection of important blocks of information used by workers such as Lamb (1977) and Flohn (1950). For example, work by Ogilvie (1991) demonstrates that a major proportion of the (Bergthórsson, 1969) reconstruction of temperature in Iceland is based on Thoroddsen's 1916 compilation which contains 'no reliable climatic information' (Ingram *et al.*, 1978, p. 331) for 24 out of 59 decades between A.D. 950 and 1540. Many of the decades with information have very little documentation, for example, only one reliable item between A.D. 1431 and 1560. Similarly, it has been demonstrated (Bell and Ogilvie, 1978; Alexandre, 1987) that a high proportion of data in climate catalogs much used by mid-twentieth century climate historians of the Middle Ages such as Lamb, for example those of Vanderlinden (1924), Easton (1928), and Hennig (1904), fail the tests of modern historical scholarship (45%, 50% and 50% of relevant material, respectively, being rejected for Vanderlinden, Easton and Hennig). A significant proportion of these errors is attributed to the uncritical re-use of second-hand or worse sources, with no direct reference to or analysis of primary sources (Bell and Ogilvie, 1978; Alexandre, 1987). Alexandre notes (1987, p. 17), for example, "innumerable errors of dating and localization which are scattered across Easton's published work". Ingram *et al.* (1978) point out that, while the correlation between Lamb's summer wetness index for England and data derived independently from the Winchester manorial rolls (Titow, 1970) is 0.77 ($n = 15$) for the decades A.D. 1200–1350, Lamb's winter severity index for England has a correlation of only 0.2 ($n = 30$) for the period A.D. 1100–1400 with Alexandre's index based on primary data from mainland Europe.

5.2. *Evidence for a Medieval Warm Epoch*

Recent work based on the rigorous application of modern techniques of historical analysis yields a somewhat different picture of European climate in the Middle Ages to that described by Lamb. For example, for the region of western and central Europe (excluding the British Isles) that he has studied, Alexandre's (1987, p. 808) conclusion is that, "the 'Medieval climate optimum' was not reached around 1150–1200, but around 1300." Ogilvie (1991, p. 249) in her analysis of Iceland's climate reaches conclusions less at variance with those of Lamb (1977), but she points out the weakness (or even absence) of reliable specific evidence for much of what has become accepted as fact. In particular, she concludes "... there is no firm foundation of data for a so-called 'Climatic Optimum' during the early years of Iceland's history. However, there is considerable circumstantial evidence for generally mild conditions up to the late twelfth century. The climate of that time may have been similar to the relatively mild climate experienced in Iceland during much of the twentieth century. From the latter part of the twelfth century, on to the sixteenth century, relatively short periods of harsh climate occurred periodically. During these times, mean annual temperatures may have fallen to 1 or 2 °C below typical twentieth century Icelandic temperatures." She points out that even these rather extreme departures cannot reasonably be seen as the onset of a 'Little Ice Age' because a prolonged cold period was not experienced in Iceland until the mid-eighteenth century, lasting until the 1840s. These recent authors have produced reconstructions of several different kinds of climate indices. For example, Alexandre (1987) has adopted a similar procedure to Lamb's (1977) in calculating an index of winter severity for all Europe north of the Alps, and indices of summer and winter precipitation for western Europe. Specific indices were developed for the Mediterranean regions (winter and summer), and similarly for spring temperatures in western Europe. He drew the following major conclusions for the various regions into which he had divided Europe on the basis of an analysis of modern instrumental data (pp. 807–808):

For Europe north of the Alps: "an important but short-lived increase of temperatures around the middle of the fourteenth century; from 1150 to 1330 a cold winter climate predominated, with more severe phases in the mid-12th century, at the beginning of the thirteenth century and in the first quarter of the fourteenth century; winter precipitation was higher from 1240 to 1400 than from 1170 to 1240."

For the mediterranean zone: "no warming in the mid-14th century."

For western Europe: "a remarkable preponderance of dry summers is established for the period 1200 to 1310, in contrast to phases of high precipitation in the second half of the twelfth century and the first half of the fourteenth century. No particular correlation is evident between these fluctuations and those of summer climate in the mediterranean region. Spring temperatures in western Europe show a long warm period from 1220 to 1310 followed by a spectacular cooling in the fourteenth century, culminating in about 1340–1350."

At least one attempt (Guiot, 1992) has been made to combine European documentary evidence with tree-rings and ice-core data. The only documentary record used that spans a major part of the Middle Ages is the decennial estimates of temperature in Iceland, but only for the decades from A.D. 1170 to 1450; the other records covering the first part of the millennium are tree-rings for various parts of Europe and Morocco and oxygen isotope data from two Greenland ice cores, in the form of quasidecadal or even longer term means. Focussing on the region between 35° and 55° N and 10° W to 20° E, Guiot (1992) reconstructs a very cold twelfth century (circa 0.5 °C below the 1851–1979 period), values close to modern from 1200 to 1300, and between modern and 0.4 degrees below modern in the fourteenth century, with the southwest of the region being the area in which the positive anomalies were greatest. The coldest episode (especially in the northeastern half of the region) was in the mid-fifteenth century, followed by the most extended warm period from the late fifteenth century to the mid-sixteenth century.

In recent years a number of important records based on historical documents from the Far East have been introduced to the scientific literature. Of these, the Chinese 'phenological temperature' record developed by Chu (1973), the frequency of dust rains, also in China (Zhang, 1984), a record of Chinese drought/flood variations (Wang *et al.*, 1987), and a drought/flood record for the region around the ancient city of Xian (Chang-an) (Li and Quan, 1987) extend through all or most of the period of interest. Just as in Europe, the acceptance and interpretation of historical documents for use in climate history requires rigorous checking of primary sources, particularly regarding the dates and locations ascribed to meteorologically related events, as well as content analysis of the texts (Zhang, 1988; Zhang and Crowley, 1989). It is not clear to the authors of this review how closely the critical methods used in the Far East resemble those in use in studying the climate history of Europe. That critical review is being conducted is demonstrated by the finding by Zhang (this volume) that the cold period described for the period around A.D. 1200 by Chu (1973) is based on a mistaken calendar conversion applied to dates of spring snowfall in Hangzhou. Zhang (ibid) also produces evidence from documentary records of cultivation of citrus and *Boehmeria nivea* that the thirteenth century cultivation limits of these plants was the northernmost in the last 1000 years, although it should be borne in mind that the distribution maps she draws are for the eighth, twelfth and thirteenth centuries specifically. She concludes that annual mean temperature and monthly mean temperature in January in south Henan province in the thirteenth century was respectively 0.9 to 1 °C and 0.6 °C higher than at present.

Li and Quan (1987) report a wet/dry index for Xian for 1604 years. This was produced in a similar manner to the indices developed for all China (State Meteorological Administration, 1981). The authors divide the whole record into episodes after spectral analysis. Within the period of interest here, 1140–1220 was reported as rather dry (one of three driest with 460–500 and 1420–1490), and 1230–1250 and 1380–1410 wet (along with 690–750, 1640–1700, and 1810–present). This evidence is suggestive of a stronger south Asian monsoon during at least some portions of the Medieval Period. In a recent paper, Sukamar *et al.* (1993) use $\delta^{13}C$ ratios to infer precipitation levels in southwestern India, concluding that a short wet phase occurred about 600 to 700 years ago. This is also consistent with a reconstruction of Nile flood deficits near Cairo by Quinn (1992), which shows that during the Medieval Period, there were few episodes of low Nile floods. This is considered a proxy record for precipitation in the Ethiopian Highlands at the headwaters of the Blue Nile.

Wang *et al.* (1987) extend the wetness/dryness index (State Meteorological Administration, 1981) back from A.D. 1470 to A.D. 950, but for 10 regions, rather than stations as for the last 500 years, because of the relative dearth of stations in this earlier period. They summarize their results in indices describing the spatial distribution of wetness and dryness between the ten regions. The eleventh and especially the twelfth centuries are reported as having a relatively high incidence of drought over the whole country as compared to 1550 to 1750, when floods over the entire country were more common. They assert that, "it is well known that most of the summers with serious floods along the Chiangjiang River are cold", and from this the conclusion might be drawn that the eleventh and twelfth centuries had fewer such cold summers in the Chiangjiang valley than the period 1550 to 1750.

5.3. *Summary of Historical Evidence*

The documentary evidence presented by authors such as Lamb for a widespread *Medieval Warm Period* has been weakened to a significant degree by recent work (Alexandre, 1988; Ogilvie, 1991) based on the rigorous application of historical criticism to sources. The remaining documentary and other historical evidence for a warm climatic period in medieval times is limited largely to evidence of crops being planted and maintained farther north or at higher elevation than in some (variable) part of the twentieth century, and to reasonable inferences based on the existence of the Iceland and Greenland settlements and some archeological evidence from those settlements. No substitute for Lamb's winter severity index, and his England annual temperature series, has yet been produced for the whole millennium. The records based on critically reviewed primary sources do reveal the existence of clusters of anomalous seasons sufficient to create decade or even perhaps multi-decade departures above or below the mean for particular seasons' temperature or precipitation.

6. Glacial-Geological Evidence

6.1. *Strengths and Limitations*

The major source of geomorphological evidence with potential to reveal climate fluctuations on time scales of centuries or less is produced by the expansion and retreat of montane glaciers, and in particular the formation of moraines. There are three major types of uncertainty associated with the use of this kind of evidence. First, there are uncertainties in dating exactly when events in moraine formation occurred that are rarely less than several decades within the period of interest in this review, and may be considerably greater. Superimposed on these uncertainties, the second problem concerns the unknown lapse of time between a change in climatic conditions and the response of the glacier as revealed in moraine formation. The third category of uncertainties arise from the complexities of the mechanisms determining glacier behavior. The first type of uncertainty – dating – receives very thorough treatment by Grove and Switsur (this volume), who also discuss problems of interpretation arising from the very uneven geographical distribution of well dated glacial geological evidence. Williams and Wigley (1983, p. 287) point out that, although glacier fluctuations may be caused by snowfall variations, or non-climatically induced glacier surges, "many historical glacier variations can be associated with measured summer temperature variations, or with a combination of summer temperature and annual precipitation changes."

6.2. *Evidence for a Medieval Warm Epoch*

Grove and Switsur (*op. cit.*) review glacial geological evidence from the European Alps, Scandinavia, the Himalaya, Alaska, the Canadian Cordillera, tropical South America, extra-tropical South America, and New Zealand. They conclude that the Medieval Warm Epoch was a global event occurring between about A.D. 900 and 1250, possibly interrupted by a minor re-advance of ice between about A.D. 1050 and 1150. The Medieval Warm Epoch is bracketed by glacial advances around the seventh to ninth centuries and the late thirteenth century, although the reliability with which these advances are dated varies greatly from one region to another, as well as within regions. In the Canadian Rockies, Luckman (this volume) reports that at least three glaciers advanced over mature forest in valley floor sites during the thirteenth century. These forests were growing 0.5 to 1 km upvalley of the most extreme recent advance of the glaciers seen in the eighteenth and nineteenth centuries, and have only been revealed by glacial retreat during the twentieth century. In northern Patagonia, Villalba (this volume) reports two major glacial advances, from A.D. 1280 to 1450, and from 1770 to 1820, and Mercer (1968, 1970) dates the initiation of major glacial advance at around 1300 in southern Patagonia. Villalba (*op. cit.*) points out that glacier advance may depend as much on increasing snowfall as on cool summers in the dry Andes of Argentina and central Chile between 30° and 35° S.

6.3. *Summary of Glacial Evidence*

There is evidence for montane glacier advances in Europe and some other regions just before A.D. 900 and after 1250, and a lack of such advances between those dates. Indeed, there is evidence of considerable glacier retreat in this period in some regions, for example, the Canadian Rockies. This general pattern is clearest and best studied in the European Alps. The record of glacial advance and particularly of retreat is much less complete in most other parts of the world, but, so far as it goes, it is not strongly inconsistent with the European pattern. To the extent that glacial retreat is associated with warm summers, the glacial geology evidence would be consistent with a warmer period in A.D. 900–1250 than immediately before or for most of the following seven hundred years.

7. Borehole Temperature Evidence

7.1. *Strengths and Limitations*

It is possible, under certain circumstances, to derive a history of ground surface temperature from measurements of the temperature-depth profile in boreholes sunk in rock, permafrost or ice. This requires an absence of effects such as ground-water movements and changes in terrain, such as surface characteristics, that would violate the assumptions on which the model used to derive the histories is based. A number of different methods of deriving temperature histories are under investigation (Beck *et al.*, 1992; Shen and Beck, 1992), which may produce a variety of outcomes. In general, since the ground acts as a low pass filter of increasing effectiveness with increasing depth, borehole-derived temperature histories have more detail for the last 100 to 200 years than for earlier periods, although the estimates for recent times may be intrinsically less stable. In particular, only the most general kind of indication of conditions in the first part of this millennium as compared to a long-term mean is given by work currently in print.

Borehole-derived annual temperature histories do have the advantage of being a direct record of past conditions not modified by human or ecological processes, providing the physical characteristics of the vegetation cover have remained constant. They may also be developed for many important regions of the Earth for which few other records of recent centuries exist, for example, much of Africa.

7.2. *Evidence for a Medieval Warm Epoch*

In most cases, exemplified by the work of Wang *et al.* (1992), the authors of borehole studies do not draw conclusions about conditions more than about 200 years ago. Referring to their work based on 23 boreholes in central and eastern Canada, Wang *et al.* (1992, p. 137) state "Due to the decreasing resolution prior to 200 years, we cannot tell from our borehole data whether there was a Little Ice Age before the cold period near the end of the last century." Very deep boreholes (several hundred meters deep) have been used by some workers to construct temperature histories of a thousand years or more. These include Beltrami *et al.* (1992) in eastern and central Canada and Mareschal and Vasseur (1992) in France. The two deep boreholes in Canada yield records with maxima in the fifteenth to sixteenth century in one case, and in the twentieth century followed by the second half of the first millennium A.D. in the other. A ground temperature history for central France was calculated using temperature-depth profiles from two boreholes. This has a maximum extending over some centuries just before and after A.D. 1000, with the amplitude of change over the past 1000 years being on the order of 1.5 °C.

7.3. *Summary*

The development of ground temperature histories from borehole temperature-depth profiles is a technique with considerable potential for providing a temporally coarse record of temperature. It has the advantages of providing a direct estimate in temperature units, and of not being restricted to recording only part of the year. There are as yet too few records from sufficiently deep boreholes to cast much light on whether there was a Medieval Warm Epoch, other than that a handful of deep boreholes yield support for a period of unknown length late in the first millennium A.D., or in the first half of the present millennium, that had annual temperatures above the long-term mean.

8. Concluding Remarks

We come back to our original question: was there a 'Medieval Warm Period', and if so, where and when? Some of the evidence compiled here and in the twelve articles of this special volume suggests that the time interval known as the *Medieval Warm Period* from the ninth to perhaps the mid-fifteenth century A.D. may have been associated with warmer conditions than those prevailing over most of the next five centuries (including the twentieth century), at least during some seasons of the year in some regions. It is obvious, however, that we have only, at best, a rough picture of the climate of this epoch, and that much work remains to be done to portray in greater detail the climatic essence of the ninth through fourteenth centuries. In particular, the simplified representations of the course of global temperature variation over the last thousand years reproduced in various technical and popular publications (for example, Eddy *et al.*, 1991; Firor, 1990; Houghton *et al.*, 1990; Mayewski *et al.*, 1993) should be disregarded, since they are based on inadequate data that have, in many cases, been superseded. Equally obvious is the fact that temporal changes displayed by nearly all of the long-term paleotemperature records examined here (see Figure 4) indicate that substantial decadal to multidecadal scale variability is present in regional temperature over the last millennium. This is, in all likelihood, a characteristic of most regional climates for the last several thousand years.

What are the implications for detecting climatic changes in the near future? As we have noted earlier, the large range of natural fluctuations in climate suggests that the occurrence of a decade or two of relatively warm climate conditions over continental scale regions does not in itself provide sufficient evidence of a change (relative to an arbitrary reference mean) in the climate. Because of generally sparse data worldwide around the turn of the first millennium, it is impossible at present to conclude from the evidence gathered here that there is anything more significant than the fact that in some areas of the globe, for some part of the year, relatively warm conditions may have prevailed. This does not constitute compelling evidence for a global 'Medieval Warm Period'. Even for the so-called Little Ice Age of later centuries, a period which overlaps part of the instrumental climate record, the picture remains incomplete, a point emphasized by Jones and Bradley (1992). There has been a marked expansion of the evidence available for the last thousand years during the course of the last 10 to 15 years. The generalized behavior of the global climate of the last millennium as a Medieval Warm Period followed by a Little Ice Age, each one or more centuries long and global in extent, is no longer supported by the available evidence. In spite of the recent surge of activity in high-resolution paleoclimatology, far more high-quality records of the climate of recent millennia would be needed to provide a definitive answer to the question: was there a Medieval Warm Period and if so, where and when? We have listed a number of important attributes that new records should possess (see section 3a). In particular, it is important that the extent to which these records faithfully record multidecade and century scale variability should be well understood.

Although climate studies of the Little Ice Age benefit from more recent and complete records than the ninth through fourteenth centuries, the evidence indicates that only a few decades in the seventeenth to nineteenth centuries A.D. showed exceptionally cold conditions which were sufficiently widespread, to represent a large-scale climatic signal. The causes of such persistent periods of warmth and coldness are varied and complex. Solar variability, volcanic activity, and ocean-atmosphere interactions all represent plausible forcing mechanisms, and human modification of the earth's energy balance through changes in the land surface and atmospheric composition are seen as having an increasingly greater impact on the behavior of climate. On the basis of the evidence currently available, it would seem most useful to concentrate research in the near future on developing the maximum number of high-quality records for as much of the last 2000 years as possible, rather than focussing on what may well turn out to be inappropriate and unhelpful concepts such as global 'Medieval Warm Periods' or 'Little Ice Ages'.

A better understanding of the nature of climate fluctuation during the last 2000 years, as proposed for the PAGES project, is likely to not only inform the study of the nature and causes of climate fluctuation on decade to century time scales. It may also help improve understanding of the interactions between human societies and climate variability, for example the role of climate in the termination of the Anasazi culture in the American Southwest and the development of the Norse colonies in the North Atlantic region. This and other features of the interaction of culture with its environment present an interesting topic for continuing and future research. Climate fluctuations on these time scales are also relevant to the dynamics of ecosystems, and in particular to the role of disturbance such as fire and extreme drought in ecosystem dynamics. We hope this special volume on the *Medieval Warm Period* serves to rekindle interest in this particular time and the climate of all of the last 2000 years.

Acknowledgements

Most of the preparation of this review was done during MKH's tenure of a Visiting Fellowship at the Cooperative Institute for Research in Environmental Science at the University of Colorado, Boulder. We are grateful to the authors of each of the tree-ring based reconstructions used in these analyses for access to their published data in electronic form, and to L. G. Thompson for permission to reproduce Figure 5. Jon Eischeid provided invaluable technical support and Gregg Garfin made helpful comments on the manuscript.

References

Alexandre, P.: 1987, *Le Climat en Europe au Moyen Age*, École des Hautes Études en Sciences Sociales, Paris.

Alley, R. and Koci, B.: 1990, 'Recent Warming in Central Greenland?', *Ann. Glaciol.* **14**, 6–8.

Beck, A., Shen, P., Beltrami, H., Mareschal, J., Safanda, J., Sebagenzi, M., Vasseur, G., and Wang, K.: 1992, 'A Comparison of Five Different Analyses in the Interpretation of Five Borehole Temperature Data Sets', *Palaeogeogr., Paleoclimatol., Paleoecol.* **98** (Global and Planetary Change Section), 101–112.

Bell, W. T. and Ogilvie, A. E. J.: 1978, 'Weather Compilations as a Source of Data for the Reconstruction of European Climate during the Medieval Period', *Clim. Change* **1**, 331–348.

Beltrami, H., Jessop, A. M., and Mareschal, J.: 1992, 'Ground Temperature Histories in Eastern and Central Canada from Geothermal Measurements: Evidence of Climatic Change', *Paleogeogr., Paleoclimatol., Paleoecol.* **98** (Global and Planetary Change Section), 167–184.

Bergthorsson, P.: 1969, 'An Estimate of Drift Ice and Temperature in Iceland in 1000 Years', *Jökull* **19**, 94–101.

Boninsegna, J.: 1990, 'Santiago de Chile Winter Rainfall since 1200 as Reconstructed by Tree Rings', in Rabassa, J. (ed.), *Quaternary of South America and Antarctica Peninsula*, Vol. 6, (1988), Balkema, Rotterdam, The Netherlands, pp. 67–68.

Boninsegna, J. A.: 1992, 'South American Dendroclimatological Records', in Bradley, R. S. and Jones, P. D. (eds.), *Climate since A.D. 1500*, Routledge, London, 446–462.

Briffa, K. R., Jones, P. D., and Schweingruber, F. H.: 1988, 'Summer Temperature Patterns over Europe: A Reconstruction from 1750 A.D. Based on Maximum Latewood Density Indices of Conifers', *Quat. Res.* **30**, 36–52.

Briffa, K. R., Jones, P. D., Bartholin, T. S., Eckstein, D., Schweingruber, F. H., Karlén, W., Zetterberg, P., and Eronen, M.: 1992, 'Fennoscandian Summers from A.D. 500: Temperature Changes on Short and Long Time Scales', *Clim. Dyn.* **7**, 111–119.

Briffa, K. R., Jones, P. D., and Schweingruber, F. H.: 1992, 'Tree-Ring Density Reconstruction of Summer Temperature Patterns across Western North America since 1600', *J. Clim.* **5**, 735–754.

Bryson, R. A. and Julian, P. R.: 1962, *Proceedings of the Conference on the Climate of the Eleventh and Sixteenth Centuries, Aspen, Colorado, June 16–24, 1962*, National Center for Atmospheric Research Technical Note 63-1, Boulder, Colorado.

Chbouki, N.: 1992, *Spatio-Temporal Characteristics of Drought as Inferred from Tree-Ring Data in Morocco*, (Unpublished Ph.D. dissertation, University of Arizona, Tucson).

Chu, Ko-Chen (Coching Chu): 1973, 'A Preliminary Study on the Climatic Fluctuations during the Last 5000 Years in China', *Sci. Sinica* **16** (2), 226–256.

Conkey, L. E.: 1986, 'Red Spruce Tree-Ring Widths and Densities in Eastern North America as Indicators of Past Climate', *Quat. Res.* **26**, 232–243.

Cook, E., Bird, T., Peterson, M., Barbetti, M., Buckley, B., D'Arrigo, R., Francey, R., and Tans, P.: 1991, 'Climatic Change in Tasmania Inferred from a 1089-Year Tree-Ring Chronology of Huon Pine', *Science* **253**, 1266–1268.

Cook, E., Bird, T., Peterson, M., Barbetti, M., Buckley, B., D'Arrigo, R., and Francey, R.: 1992, 'Climatic Change over the Last Millennium in Tasmania Reconstructed from Tree-Rings', *Holocene* **2**, 205–217.

Cook, E. R., Bird, T., Peterson, M., Barbetti, M., Buckley,B., and Francey, R.: 1992, 'The Little Ice Age in Tasmanian Tree Rings', in Mikami, T. (ed.), *Proceedings of the International Symposium on the Little Ice Age*, Tokyo Metropolitan University, pp. 11–17.

Dansgaard, W., Johnsen, S., Reeh, N., Gundestrup, N., Clausen, H., and Hammer, C.: 1975, 'Climatic Changes, Norsemen and Modern Man', *Nature* **255**, 24–28.

D'Arrigo, R. D. and Jacoby, G. C.: 1991, 'A 1000-Year Record of Winter Precipitation from Northwestern New Mexico, U.S.A.: A Reconstruction from Tree-Rings and Its Relation to El Niño and the Southern Oscillation', *Holocene* **1, 2**, 95–101.

Diaz, H. F. and Bradley, R. S.: in press, 'Documenting Natural Climatic Variations: How Different is the Climate of the Twentieth Century from that of Previous Centuries?', in *Climate Variability on Decade to Century Time Scales*, Washington, D.C., National Academy of Science.

Duchon, C. E.: 1979, 'Lanczos Filtering in One and Two Dimensions', *J. Appl. Meteor.* **18**, 1016–1022.

Easton, C.: 1928, *Les Hivers dans l'Europe Occidentale*, Brill, Leyden.

Eddy, J. A., Bradley, R. S., and Webb, T.: 1991, 'Earthquest', *OIES Publ.* **5 (1)**, University Corporation for Atmospheric Research, Boulder, Colorado.

Firor, J.: 1990, *The Changing Atmosphere: A Global Challenge*, Yale University Press, New Haven.

Flohn, H.: 1950, 'Klimaschwankungen im Mittelalter und Ihre Historisch-Geographische Bedeutung', *Berichte Landesk.* **7**, 347–357.

Fritts, H. C.: 1991, *Reconstructing Large-Scale Climatic Patterns from Tree-Ring Data*, The University of Arizona Press, Tucson.

Gottschalk, M. K. E.: 1971–1977, *Stormvloeden en Rivieroverstromingen in Nederland*, Assen, (3 vols.).

Graumlich, L. J.: 1993, 'A 1000-Year Record of Temperature and Precipitation in the Sierra Nevada', *Quatern. Res.* **39**, 249–255.

Graybill, D. A.: 1993, 'Dendroclimatic Reconstructions during the Past Millennium in the Southern Sierra Nevada and Owens Valley, California', in Lavenberg, R. (ed.), *Southern California Climate: Trends and Extremes of the Past 2000 Years*, Natural History Museum of Los Angeles County, Los Angeles, CA, pp. 1–24.

Graybill, D. A. and Idso, S. B.: 1993, 'Detecting the Aerial Fertilization Effect of Atmospheric CO_2, Enrichment in Tree-Ring Chronologies', *Glob. Biogeochem. Cycl.* **7**, 81–95.

Graybill, D. A. and Shiyatov, S. G.: 1989, 'A 1009 Year Tree-Ring Reconstruction of Mean June–July Temperature Deviations in the Polar Urals', in White *et al.* (ed.), *Air Pollution Effects on Vegetation including Forest Ecosystems*, pp. 37–42.

Graybill, D. A. and Shiyatov, S. G.: 1992, 'Dendroclimatic Evidence from the Northern Soviet Union', in Bradley, R. S. and Jones, P. D. (eds.), *Climate since A.D. 1500*, Routledge, London, pp. 393–414.

Guiot, J., Tessier, L., Serre-Bachet, F., Guibal, F., and Gadbin, C.: 1988, 'Annual Temperature Changes Reconstructed in W. Europe and N.W. Africa back to A.D. 100', *Annal. Geophys.* **85**, (Special Issue, XIII General Assembly of CGS, Bologne).

Guiot, J.: 1992, 'The Combination of Historical Documents and Biological Data in the Reconstruction of Climate Variations in Space and Time', in Frenzel, B., Pfister, C., and Gläser, B. (eds.), *European Climate Reconstructed from Documentary Data: Methods and Results. Special Issue 2 of the Journal Paleoclimate Research, Volume 7*, Akademie der Wissenschaften und der Literatur, Mainz; Stuttgart; Jena; New York, pp. 93–104.

Hennig, R.: 1904, 'Katalog Bemerkenswerter Witterungsereignisse von den Ältesten Zeiten bis zum Jahre 1800', *Abhandl. Preuss. Akad. Met. Inst.* **2** (4).

Houghton, J. T., Jenkins, G. J., and Ephraum, J. J. (eds.): 1990, *Climate Change: The IPCC Scientific Assessment*, Cambridge University Press, Cambridge.

Hughes, M. K.: 1992, 'Dendroclimatic Evidence from the Western Himalaya', in Bradley, R. S. and Jones, P. D. (eds.), *Climate since A.D. 1500*, Routledge, London and New York, pp. 415–431.

Hughes, M. K. and Brown, P. M.: 1992, 'Drought Frequency in Central California since 101 B.C. Recorded in Giant Sequoia Tree Rings', *Clim. Dyn.* **6**, 161–167.

Hughes, M. K. and Davies, A. C.: 1987, 'Dendroclimatology in Kashmir Using Tree Ring Widths and Densities in Subalpine Conifers', in Kairiukstis, L., Bednarz, Z., and Feliksik, E. (eds.), *Methods of Dendrochronology: East-West Approaches*, (IIASA/PAN, Vienna), pp. 163–176.

Hughes, M. K., Schweingruber, F. H., Cartwright, D., and Kelly, P. M.: 1984, 'July–August Temperature at Edinburgh between 1721 and 1975 from Tree-Ring Density and Width Data', *Nature* **308**, 341–344.

Hughes, M. K., Richards, B. F., Swetnam, T. W., and Baisan, C. H.: 1990, 'Can a Climate Record Be

Extracted from Giant Sequoia Tree Rings?', in Betancourt, J. and MacKay, A. (eds.), *Proceedings of the Sixth Annual Pacific Climate (PACLIM) Workshop*.

Ingram, M. J., Underhill, D. J., and Wigley, T. M. L.: 1978, 'Historical Climatology', *Nature* **276**, 329–334.

Ingram, M. J., Underhill, D. J., and Farmer, G.: 1981, 'The Use of Documentary Sources for the Study of Past Climates', in Wigley, T. M. L., Ingram, M. J., and Farmer, G. (eds.), *Climate and History*, Cambridge University Press, pp. 180–213.

International Geosphere-Biosphere Program/International Council of Scientific Unions: 1990, 'Global Change', Report No. 12, Chapter 7.

Jones, P. D. and Bradley, R. S.: 1992, 'Climatic Variations over the Last 500 Years', in Bradley, R. S. and Jones, P. D. (eds.), *Climate since A.D. 1500*, Routledge, London, pp. 649–665.

Jones, P. D., Raper, S. C. B., Bradley, R. S., Diaz, H. F., Kelly, P. M. and Wigley, T. M. L.: 1986, 'Northern Hemisphere Surface Air Temperature Variations: 1851–1984', *Journal of Climate and Applied Meteorology* **25**, 161–179.

Jones, P. D., Marsh, R., Wigley, T., and Peel, D.: 1993, 'Decadal Time Scale Links between Antarctic Peninsula Ice-Core Oxygen-18, Deuterium and Temperature', *Holocene* **3**, 14–26.

Ladurie, E. L.: 1971, *Times of Feast, Times of Famine*, Doubleday, New York.

LaMarche, V. C., Jr.: 1974, 'Paleoclimatic Inferences from Long Tree-Ring Records', *Science* **183**, 1043–1048.

Lamb, H. H.: 1965, 'The Early Medieval Warm Epoch and Its Sequel', *Palaeogeogr., Paleoclimatol., Palaeoecol.* **1**, 13–37.

Lamb, H. H.: 1977, *Climatic History and the Future*, Vol. 2: *Climate: Present, Past and Future*, Methuen and Co. Ltd., London, England.

Lamb, H. H.: 1982, *Climate, History and the Modern World*, Methuen and Co. Ltd., London.

Lara, A. and Villalba, R.: 1993, 'A 3620-Year Temperature Record from *Fitzroya cupressoides* Tree Rings in Southern South America', *Science* **260**, 1104–1106.

Li, Zhaoyan and Quan, Xiaowei: 1987, 'The Climatic Changes of Drought-Wet in Ancient Chang-an Region of China during the Last 1604 Years', in Ye Duzheng, Fu Congbin, Chao Jiping, and M. Yoshino (eds.), *The Climate of China and Global Climate*, Springer-Verlag, Berlin, 1987, pp. 57–62.

Mayewski, P. A., Meeker, L. D., Morrison, M. C., Twickler, M. S., Whitlow, S. I., Ferland, K. K., Meese, D. A., Legrand, M. R., and Steffensen, J. P.: 1993, 'Greenland Ice Core "Signal" Characteristics: An Expanded View of Climate Change', *J. Geophys. Res.* **98**, 12 839–12 847.

Mareschal, J. and Vasseur, G.: 1992, 'Ground Temperature History from Two Deep Boreholes in Central France', *Palaeogeogr., Paleoclimatol., Palaeoecol.* **98** (Global and Planetary Change Section), 185–192.

Mercer, J. H.: 1968, 'Variations of Some Patagonian Glaciers since the Late Glacial', *Amer. J. Sci.* **266**, 91–109.

Oeschger, H. and Langway, C. (eds.): 1989, *The Environmental Record in Glaciers and Ice Sheets*, John Wiley and Sons, Chichester.

Oeschger, H. and Arquit, A.: 1991, 'Resolving Abrupt and High-Frequency Global Changes in the Ice-Core Record', in Bradley, R. (ed.), *Global Changes of the Past*, Bolder, CO: UCAR/Office of Interdisciplinary Earth Studies, pp. 175–200.

Ogilvie, A. E. J.: 1991, 'Climatic Changes in Iceland, A.d. 865 to 1598', in *The Norse of the North Atlantic* (Presented by G. F. Bigelow), *Acta Archaeol.* **61**, Munksgaard, Copenhagen, pp. 233–251.

Ogilvie, A. E. J.: in press, 'Documentary Records of Climate from Iceland during the Late Maunder Minimum Period A.D. 1675 to 1715 with Reference to the Isotopic Record from Greenland', *Päleoklimaforschung*.

Pfister, C.: 1988, 'Variations in the Spring–Summer Climate of Central Europe from the High Middle Ages to 1850', in Wanner, H. and Siegenthaler, U. (eds.), *Long and Short Term Variability of Climate*, Berlin, Springer Verlag, pp. 57–82.

Quinn, W.: 1992, 'A Study of Southern Oscillation-Related Climatic Activity for A.D. 622–1990 Incorporating Nile River Flood Data', in Diaz, H. F. and Markgraf, V. (eds.), *El Niño: Historical and Paleoclimatic Aspects of the Southern Oscillation*, Cambridge University Press, Cambridge, pp. 119–149.

Robin, G. D. Q. (ed.): 1981, *The Climatic Record in Polar Ice Sheets*, Cambridge University Press, Cambridge.

Rose, M. R., Dean, J. S., and Robinson, W. J.: 1981, *The Past Climate of Arroyo Hondo, New Mexico, Reconstructed from Tree Rings*, Arroyo Hondo Archeological Series, Vol. 4, School of American Research Press, Santa Fe, New Mexico.

Scuderi, L. A.: 1993, 'A 2000-Year Tree ring Record of Annual Temperatures in the Sierra Nevada Mountains', *Science* **259**, 1433–1436.

Shen, P. and Beck, A.: 1992, 'Paleoclimate Change and Heat Flow Density Inferred from Temperature Data in the Superior Province of the Canadian Shield', *Palaeogeogr., Palaeoclimatol., Palaeoecol.* **98** (Global and Planetary Change Section), pp. 143–165.

Stahle, D. W. and Cleaveland, M. K.: 1992, 'Reconstruction and Analysis of Spring Rainfall over the Southeastern U.S. for the Past 1000 Years', *Bull. Amer. Meteorol. Soc.* **73**, 1947–161.

Stahle, D. W., Cleaveland, M. K., and Hehr, J. G.: 1985, 'A 450-Year Drought Reconstruction for Arkansas, United States', *Nature* **316**, 530–532.

State Meteorological Administration: 1981, 'Annals of 510 Years of Precipitation Records in China', Meteorological Research Institute, Beijing.

Sukumar, R., Ramesh, R., Pant, R. K., and Rajagopalan, G.: 1993, 'A δ^{13}C Record of Late Quaternary Climate from Tropical Peats in Southern India', *Nature* **364**, 703–706.

Taylor, K., Rose, M., and Lamorey, G.: 1992, 'Relationship of Solar Activity and Climatic Oscillations on the Colorado Plateau', *J. Geophys. Res.* **97**, 803–15, 811.

Thompson, L.: 1991, 'Ice-Core Records with Emphasis on the Global Record of the Last 2000 Years', in Bradley, R. (ed.), *Global Changes of the Past*, Boulder, CO: UCAR/Office for Interdisciplinary Earth Studies, pp. 201–224.

Thompson, L. and Mosley-Thompson, E.: 1989, 'One-Half Millennia of Tropical Climate Variability as Recorded in the Stratigraphy of the Quelccaya Ice Cap, Peru', in Peterson, D. (ed.), *Aspects of Climate Variability in the Pacific and the Western Americas, Geophysical Monograph 55*, Washington, DC, American Geophysical Union, pp. 15–31.

Till, C. and Guiot, J.: 1990, 'Reconstruction of Precipitation in Morocco since 1100 A.D. Based on Cedrus atlantica Tree-Ring Widths', *Quat. Res.* **33**, 337–351.

Titow, J.: 1970, 'Le Climat à travers les rôles de comptabilité de l'Évêché de Winchester (1350–1450)', *Annal. E.S.C.* **25**, 312–350.

U.S. Dept. of Energy: 1989, *Atmospheric Carbon Dioxide and the Greenhouse Effect*, Available from NTIS, Springfield, VA, 36 pp.

Vanderlinden, E.: 1924, 'Chronique des Événements Météorologiques en Belgique jusqu'en 1834', *Mém. Acad. Roy. Belge* **5** (2).

Villalba, R.: 1990, 'Climatic Fluctuations in Northern Patagonia in the Last 1000 Years as Inferred from Tree-Ring Records', *Quat. Res.* **34**, 346–360.

Villalba, R., Leiva, J. C., Rubullis, S., Suarez, J., and Lenzano, L.: 1990, 'Climate, Tree-Ring, and Glacial Fluctuations in the Río Frías Valley, Rio Negro, Argentina', *Arc. Alp. Res.* **22**, 215–232.

Wang, K., Lewis, T. J., and Jessop, A. M.: 1992, 'Climatic Changes in Central and Eastern Canada Inferred from Deep Borehole Temperature Data', *Palaeogeogr., Palaeoclimatol., Palaeoecol.* **98** (Global and Planetary Change Section), pp. 129–141.

Wang, Shao-wu and Mearns, L. O.: 1987, 'The Impact of the 1982–83 El Niño Event on Crop Yields in China', NCAR/UNEP Report, in Glantz, M. *et al.* (eds.), *The Societal Impacts of World Wide Climate Anomalies during the 1982–83 El Niño*, (NCAR, Boulder, CO), pp. 43–49.

Wang, Shaowu, Zhao, Zongci, Chen Zhenhua, and Tang Zhongxin: 1987, 'Drought/Flood Variations for the Last 2000 Years in China and Comparison with Global Climatic Change', in Ye Duzheng, Fu Congbin, Chao Jiping, and M. Yoshino (eds.), *The Climate of China and Global Climate*, Springer-Verlag, Berlin, 1987, pp. 20–29.

Williams, L. D. and Wigley, T. M. L.: 1983, 'A Comparison of Evidence for the Late Holocene Summer Temperature Variations in the Northern Hemisphere', *Quat. Res.* **20**, 286–307.

Zhang, De'er: 1984, 'Synoptic-Climatic Studies of Dust Fall in China since Historic Times', *Sci. Sin.* **B27**, 825–836.

Zhang, De'er: 1988, 'The Method for Reconstruction of the Dryness/Wetness Series in China for the

Last 500 Years and Its Reliability', in Zhang, J. (ed.), *The Reconstruction of Climate in China for Historical Times*, Science Press, Beijing, China, pp. 18–31.
Zhang, J. and Crowley, T. J.: 1989, 'Historical Climate Records in China and Reconstruction of Past Centuries', *J. Clim.* **2**, 833–849.

(Received 22 September, 1993; in revised form 9 December, 1993)

GLACIAL GEOLOGICAL EVIDENCE FOR THE MEDIEVAL WARM PERIOD

JEAN M. GROVE

Girton College, Cambridge, U.K.

and

ROY SWITSUR

Wolfson College, Cambridge, U.K.

Abstract. It is hypothesised that the Medieval Warm Period was preceded and followed by periods of moraine deposition associated with glacier expansion. Improvements in the methodology of radiocarbon calibration make it possible to convert radiocarbon ages to calendar dates with greater precision than was previously possible. Dating of organic material closely associated with moraines in many montane regions has reached the point where it is possible to survey available information concerning the timing of the medieval warm period. The results suggest that it was a global event occurring between about 900 and 1250 A.D., possibly interrupted by a minor readvance of ice between about 1050 and 1150 A.D.

1. Introduction

During the last few hundred years glaciers in both hemispheres expanded. Their maximal positions can be identified by the moraines they left behind a kilometre or two away from the present retreating ice fronts. Moraines formed by earlier periods of enlargement can, given favourable circumstances, be dated. Dating of organic material above a moraine provides a minimal age for its formation. Material from beneath it can provide a maximum age. A sample from within a moraine may give a closer approximation of the time of formation. Time lags almost always occur between moraine formation and the accumulation or deposition of the dated material.

It can reasonably be hypothesised that the Medieval Warm Period (MWP) should be bracketed by the dates of the moraines formed during the periods of enlargement which preceded and followed it. In Europe and in some other parts of the world glacial expansion is known to have taken place between 800 A.D. and 900 A.D. There is at present no commonly accepted name for this episode. The most recent period of glacial expansion, during which glaciers remained more advanced than in the mid to late twentieth century, occurred in the six centuries between about 1250–1300 A.D. and 1850–90 A.D. The last four centuries of this period are widely known as the Little Ice Age (LIA). Some find it more appropriate to apply the term LIA to the whole period from the thirteenth to the nineteenth centuries, especially as although it covered two advance phases of approximately

Climatic Change **26**: 143–169, 1994.
© 1994 *Kluwer Academic Publishers.*

equal magnitude, conditions between them did not revert to those obtaining before 1300 A.D. In this paper the broader definition of the LIA, covering 1250–1300 A.D. to 1850–90 A.D., is adopted.

Recently some doubt has been cast on the utility of retaining the term LIA (Bradley and Jones, 1992a, b), because recent centuries are now known to have been characterised by 'complex climatic anomalies, with both warm and cold episodes which varied in importance geographically' (Bradley and Jones, 1993), changes in the general circulation resulting in positive anomalies in some areas at times when negative anomalies are recorded elsewhere. However when the many individual proxy climatic records brought together in 'Climate since A.D. 1500' (Bradley and Jones 1992b) are combined, it is revealed that in the northern hemisphere temperatures from 1400 A.D. to 1910 were between 0.5 and 1.5 °C below those experienced during the twentieth century, except for the last quarter of the fifteenth century, the first half of the sixteenth century and the latter half of the eighteenth century (Bradley and Jones, 1993). (The southern hemisphere proxy data presently available is insufficient to support generalisation.) This is very much what was to be expected from the glacial evidence, and justifies the continued use of the term LIA to describe a period during which the fronts of enlarged glaciers oscillated around forward positions.

In practice it is often possible to obtain only a maximum or only a minimum date for the formation of a particular moraine, but by now a considerable amount of information is none the less available. It is intended here to survey the existing evidence which should, at least in theory, serve to reveal whether the MWP was global in extent, and whether it occurred at the same time in different regions.

Certain difficulties and limitations occur in practice, firstly with regard to dating. Apart from very recent features, about which historical data is available, this depends on dendrochronology, lichenometry or radiocarbon. Obtaining the age of the oldest trees or tree growing on a moraine may well not provide a minimal age anywhere near the real age. The oldest tree may not be found. It may anyway not belong to the first generation to grow on the moraine. The unknown lag between moraine formation and tree growth may be very long indeed.

2. Dating Methods

2.1. *Lichenometry*

Lichenometry is dependent on the hypothesis that the largest lichen growing on a surface is the oldest, and that if the growth rate for a given species is known, the maximum lichen size will give a minimum age for the substrate, the rate being unaffected by changes in climate. Lichenometry can be accurate to within ±5 years or so for the last 200 years, given unusually favourable circumstances (Porter, 1981). Acceptable accuracy may be obtained for considerably longer periods, but lichenometry cannot be considered useful for the MWP. Studies based on lichenometric dating will not be discussed in this paper.

2.2. Radiocarbon Dating

The use of radiocarbon analysis depends upon the existence and discovery of suitable materials for dating. Carbon dating of mature paleosols involves the possibility of substantial errors (Matthews, 1984, 1985). If young soils are dated, separation of the humic acids from the humins, allows the necessary evaluation of the extent of contamination. It is suggested that the humic acid fraction may be taken to give the younger age, representing the end of soil formation. Macro-residuals, such as lichens, may give an approximation of the beginning of soil formation, following an earlier cold phase (Röthlisberger and Geyh, 1985a).

There are also potential difficulties with dating peat of any appreciable age. In the upper few centimetres, the *acrotelm*, the peat has an open structure and rapid and almost complete oxidation occurs, but in the lower, more compacted *catatelm* the breakdown becomes very slow in the prevailing conditions of oxygen deficiency. Decomposition may take place under these conditions through the agency of sulphur-fixing bacteria, hence giving rise to the characteristic fetid smell of such deposits. The decomposition products include complex fulvic and humic acids of various high molecular weights, and since the reactions do not proceed to completion, isotopic fractionation takes place. Slow compaction of the peaty residue accompanies the chemical changes. The residual humins represent a peat fraction that is more resistant to decomposition. Since both humic acids and fulvic acids are to some extent soluble forming colloidal solutions, these may be transported by water movement and possibly left as contamination of a different age elsewhere in the deposit. This is more likely to occur in deep or ancient deposits. Recent work on dating the various fractions from blanket peat has demonstrated inconsistencies with the humic acids sometimes younger and sometimes older than the humin fraction and differing in age by up to several centuries. Such age differences are much greater than might be expected from uncertainties of the technique. Hence, for moraine dating, wood, moss or thin peats would appear to provide a safer basis for dating.

3. Calibration

A problem with radiocarbon dating of Holocene samples is that radiocarbon years and sidereal years are rarely of the same length. This phenomenon is caused by the secular variation of the concentration of atmospheric radiocarbon. It was clearly demonstrated by Willis *et al.* (1960), from measurements on small groups of tree rings of known dendro-date, taken from a *Giant Redwood*, that during the Christian era the radiocarbon concentrations had not been constant, as required by the simple radiocarbon dating theory of Libby, but had varied by 1% or more, causing the radiocarbon age of samples to differ from calendar date by 50 to 100 years. Similar research by other workers on long-lived species, such as *Bristlecone Pine* (Suess, 1970), showed that the effect became more pronounced for earlier periods, and that a discrepancy of almost a millennium occurred around 6000 BP In order to obtain results compatible with calendar years it has thus become necessary to *calibrate* the radiocarbon time scale using dendrochronologically dated timber. Many calibration studies have been carried out during the past three decades and the most recent measurements have been made with higher accuracy and greater precision than those of earlier years. The latest work has invoved *oak* and *Douglass pine* and the combined results presented by Pearson, Stuiver and their co-workers have been accepted as the best possible at the present time. However the data as published (Stuiver and Pearson, 1986, 1993 and others) are not ideal for use by hand and do not allow the full information to be extracted easily, and so several attempts have been made to adapt their use to computer manipulation. A comparison of some of these different techniques was given at the Dubrovnik Radiocarbon International Conference (Aitchison *et al.*, 1989). The general agreement between the algorithms was good and the discrepances amongst the techniques were only of a few years.

Measurements of radiocarbon activity are essentially of a statistical nature, being repeated timed counts of nuclear disintegrations of a specified energy range during given time. The results of these may be represented by a distribution that has a well defined mean value and standard deviation. A similar distribution is obtained for the radioactivity of the international standard, NBS oxalic acid, which represents the activity of the modern pre-industrial revolution atmosphere. These data are further corrected for instrumental blanks and other laboratory errors and for the isotopic fractionation of the sample material. Finally the activity is converted by a logarithmic transformation to a *conventional radiocarbon age*, BP, (before present) which is based on the zero year of A.D. 1950 and the Libby half-life for the radiocarbon isotope, of 5568 years. In the recent part of the time scale, the age probability has a Gaussion distribution with a mean age and uncertainty. The uncertainty is conventionally represented by a $\pm$ term at a probability of 68%, which implies a single standard deviation of the disintegrated statistics. It has been found that this value tends to underestimate the true uncertainty of the age and to compensate for this another multiplying factor (ca. 1.1 to 2.5 for example), is often used. Unfortunately such factors differ for each radiocarbon laboratory, each instrument in that laboratory and probably with time, and this has the effect of smearing the radiocarbon age over a greater period of time. The measured, known-date calibration points, of course, are all similar Gaussian distributions and the shape of the calibration curve itself is the envelope of these distributions. In practice the curve is obtained by computer fitting a curve to the individual points, normally using a spline function.

In order to calibrate a particular conventional radiocarbon age, one technique is to divide the appropriate portion of the calendar axis into small sections of say 20 years. The probability of the true-date falling into each of these sections is evaluated by the interaction of the appropriate Gaussian distribution of the radiocarbon age with that of the calibration curve. The probability distribution for the calendar range is thus constructed on the calendar axis, conveniently in the form of a histogram.

The extreme ends of the distribution represent the calibrated range for that date. The distribution along the calendar axis is not a symmetrical Gaussian unless the calibration curve is a straight line. In general this is not the case because of the wiggles of the calibration curve.

4. Examples of Calibration

The process of calibration may be followed for different cases in Figures 1a and 1b, for two radiocarbon ages used in this survey, which show quite different effects on the date range. In these the ordinate represents the radiocarbon age (BP) axis and the abscissa is the axis for the calendar date-ranges. The internationally accepted notation for this axis is cal.A.D. (or cal.B.C. or cal.BP). The dashed straight line sloping from left to right across the diagram represents the line of equivalence of radiocarbon ages and calendar dates. The winding curve sloping downwards left to right represents the calibration curve and it will be seen how his moves about the line of equivalence. This calibration curve is intersected by three horizontal lines which represent the conventional radiocarbon age. The middle line is the central age, and the other two indicate the uncertainty equivalent to the one standard deviation limits of the statistics producing this age. The interaction between these produce the shaded histogram drawn on the calendar-date axis. The heights of the histogram rectangles are proportional to the probability (indicated on the right hand scale) of the calibrated date being at that position. In the computation the extremes of the histogram represent 95% confidence limits for the calibrated date. It is customary to indicate the conventional radiocarbon measurements by the term *age* and the calibrated version by *date-range*.

The two diagrams show well the different effects of the shape of the calibration curve as well as the age uncertainty on the calibrated date-range. In Figure 1a the intersection of the calibration curve and the radiocarbon age line occurs where the slope of the calibration curve is at a high angle to the calendar date axis. The standard deviation also is quite small, so that the probability histogram extends only a short distance along the calendar axis. This leads to the calibrated date having a *smaller time range* than the original radiocarbon age. In this instance, at 68% confidence limits, the radiocarbon age spans 90 years (765 to 675 BP), whereas the calibrated date-range is only 20 years and may be pinpointed to between 1250 and 1270 cal.A.D. By contrast however in Figure 1b the conditions are quite different; at the intersection the calibration curve is almost parallel to the calendar axis, and this, together with the large uncertainty of the age, produces a probability histogram with a wide range. The histogram has several relatively high peaks indicating multiple valid possibilities for the correct date. The 68% confidence limits here are necessarily wider, at 245 years, than the limits of the radiocarbon age, 200 years. The value of the histogram height helps to interpret the likelihood of any particular date being correct. However in our survey paper the data, from many sources, are of necessity of varying precision and reliability and do not provide the full pattern

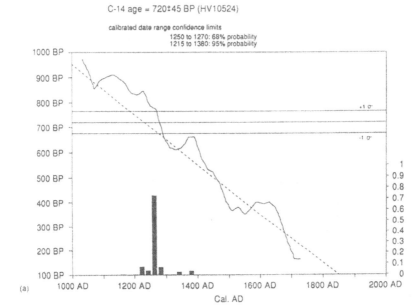

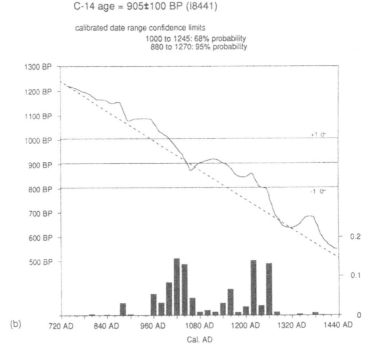

Fig. 1.   Examples of calibration of radiocarbon years to calendar years: (a) Hv-10524, a sample from a moraine of the Balfour Glacier, Westland, New Zealand; (b) I-8441, a sample from peat overridden by the Quelcaya Icecap, Peru.

of glacier frontal oscillations, so the detailed aspect of the calibration procedure will not be pursued further. The calibrated dates in the tables are given in terms of the date-range falling within the 68% (1 sigma statistics) probability range. The dated evidence relating to glacier fluctuations has accumulated from many parts of the world to an extent which now makes it worthwhile to attempt to assess its value in the context of the Medieval Warm Period.

5. The Distribution of the Data and Its Implications

Difficulties of dating are not the only ones. The major problem caused by the distribution of investigations must not be overlooked. The bulk of detailed research has been carried out in Europe. This has revealed that the commonly held view that the Little Ice Age (LIA) advances were everywhere the most extensive is not correct. In the European Alps the ice reached very similar positions several times during the Holocene, causing deposits to be set down very close to each other or actually on top of each other. Only very intensive field investigations have served to reveal the complexity of Holocene glacial history. While it should not be assumed that these relationships between glacial phases in the last few thousand years are to be expected in other regions, it certainly cannot be taken for granted that advances elsewhere were well separated spatially. Adequate investigations must therefore be intensive. But regions such as the southern Andes or the Canadian Rockies contain thousands of glaciers which have never been examined. In Alaska it is, very naturally, the major ice streams which have attracted attention, yet most of these either have tongues prone to float, or are subject to surging. Such glaciers may exhibit substantial fluctuations unconnected with minor or intermediate scale changes in climatic conditions. Although they also react to climatic circumstances (Mann, 1986) evidence from such glaciers will not be considered here. Dates which have had to be corrected for carbon deficiency in Antarctic sea water by several hundred years, such as those from the South Shetland Islands (Clapperton and Sugden, 1989) are also not discussed.

Key evidence from regions in Eurasia, North and South America, and New Zealand (see Figure 2) will be presented in the form of tabulated graphs, showing both conventional radiocarbon ages, as originally published, with their laboratory errors, and their calendar equivalents at 68% probability (Stuiver and Pearson, 1986).

6. The European Alps

The best known Holocene glacial history is that of the Alps, particularly the Swiss Alps, resulting in large part from the work of a group of researchers at the Geographical Institute of Zürich University. They had the advantage of the discovery of much fossil wood amongst the glacial deposits. (Between 1970 and 1975, thirty pieces were found within 300 m of the Zmutt glacier alone). The numbers of

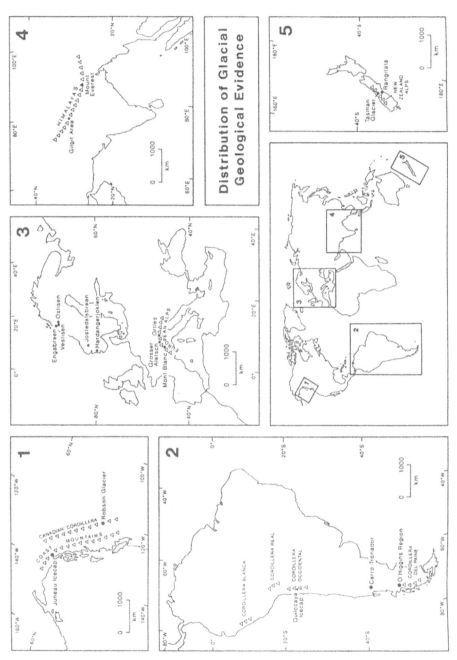

Fig. 2.   The location of glacial geological evidence discussed in the text.

rings in tree trunks overridden by ice, it was realised, could be used to give minimum values for the length of periods between advances. By 1980, Röthlisberger and his associates had obtained some 87 dates relevant to the Holocene sequence in the canton of Valais alone, indicating that the glaciers had advanced to very similar positions in Medieval times and in the LIA, but the abundance of dates around 1000±100 BP and their overlap underlined the difficulties of interpretation (Röthlisberger *et al.*, 1980), and suggested the potential importance of combining glacial and dendroclimatological evidence.

One of the earlier discoveries was that of wood fragments, which had been buried by the Aletsch Glacier (Oeschger and Röthlisberger, 1961). It was concluded that the trees had been covered by an ice advance about 1200 A.D., nearer the end of the MWP. Retreat of the tongue during the twentieth century uncovered not only the stumps of trees growing *in situ* and roots, but also the remains of an irrigation channel, the Oberriederin Bisse, which is recorded as abandoned in 1385 (Kinzl, 1932). Lamb (1965) who first advanced the idea of the MWP, argued that this bisse (a *bisse* is an irrigation channel, essential for agriculture in the dry climate of the Valais), had been constructed during the warm period in a position which became untenable when the climate changed and the glacier expanded (Lamb, 1988).

It was the Aletsch which was to furnish evidence allowing the most detailed reconstruction of glacier oscillations before, during and after the MWP. Holzhauser (1984, 1988) has made intensive studies of the Aletsch deposits and has been able to reconstruct a remarkably complete record of the fluctuations of its tongue for the last three thousand years (Table I, Figure 3). This shows that the MWP was preceded by a rapid advance of the ice, lasting about a century which, around 850–900 A.D., brought its front close to its LIA maximum. Retreat of the ice between 900 A.D. and 1000 A.D. was almost as swift. The MWP was terminated by an equally rapid advance culminating about 1350 A.D. But climatic conditions were not uniform throughout the MWP. The Aletsch readvanced again from 1050 A.D. to 1150 A.D., to a position about 1.7 km beyond the current one. The earliest phase of the LIA ended the MWP with an abrupt and substantial advance between 1250 A.D. and 1350 A.D., destroying part of the Oberriederin Bisse. Documentary evidence that the Allalin glacier was also in an extended position in the thirteenth century has also been found (Lütsch, 1926).

The Aletsch reconstruction is exceptionally accurate not only because of the wealth of radiocarbon evidence but also because from 1100 A.D. onwards it is controlled by absolute dating of tree rings from stumps *in situ*. Additional documentary sources have also been located. Confirmatory evidence of many of its features comes from the Grindelwald, the Rhone and the Unteraar glaciers (Figure 4). The oscillations of a large sample of Swiss glaciers have been observed annually since the late nineteenth century (Grove, 1988). Those of the Aletsch, Grindelwald, Rhone and Unteraar have been in no way exceptional, and their earlier fluctuations may reasonably be taken as representative for the region. Further critical reexam-

TABLE I: Key dates relating to the glacier oscillations in the Swiss, French and Italian Alps. The numbers ⁎ the margin give the laboratory identification number, the C14 age and the calibrated age for 68% probability Horizontal lines represent calibrated dates. The symbol > indicates a maximum date; < indicates a minimum date

	600	700	800	900	1000	1100	1200	1300	1400AD
SWISS ALPS (Zumbühl and Holzhauser, 1990)									
Grosser Aletsch									
W(UZ-682)1195/70/730–950		>============							
W(UZ-683)1145/70/800–970			>========						
W(B-3961)1140/70/800–975			>========						
W(UZ-516)1140/70/800–980			>========						
W(UZ-535)1055/60/885–1020				>————					
W(UZ-381)900/60/1020–1220						>============			
W(UZ-384)815/60/1095–1260						>============			
W(B-32)729/100/1210–1375								>————	
W(B-71)800/100/1045–1270						>————			
W(B-3962)780/20/1220–1265							>==		
W(UZ-383)755/60/1215–1270							>—		
W(UZ-525)700/60/1250–1340								>=====	
W(UZ-380)665/55/1215–1270							>====		
W(UZ-333)620/60/1260–1395								>======	
W(UZ-523)605/65/1265–1405								>=======	
Unteraar									
W(UZ-954)1085/75/875–1015				>————					
S(UZ-1042)945/80/990–1210					>—————				
W(UZ-963)890/75/1015–1230					>—————				
Rhone									
S(UZ-947)800/75/1085–1265						>—————			
S(UZ-2262)695/90/1225–1380							>————		
S(UZ-829)679/65/1255–1375								>———	
S(UZ-830)665/70/1255–1385								>———	
SWISS ALPS (Zumbühl and Holzhauser, 1990) cont.									
Untergrindel Wald									
W(UZ-1294)920/75/1005–1220					>—————				
W(UZ-1169)945/75/995–1190					>————				
W(UZ-1171)865/75/1030–1250					>————————				
W(UZ-1157)935/75/1000–1210					>————				
W(UZ-1161)890/75/1015–1230					>—————				
W(UZ-1158)865/75/1030–1250					>————————				
S(UZ-1151)775/75/1165–1265						>————			
Unteraar)									
S(UZ-1042)945/80/990–1210					>—————				
W(UZ-963)890/75/1015–1230					>—————				
	600	700	800	900	1000	1100	1200	1300	1400AD

TABLE I: *Continued*

	600	700	800	900	1000	1100	1200	1300	1400AD
FRENCH ALPS									
Trient (Bless, 1984)									
S(UZ-233)825/55/1065–1260						>———————			
ITALIAN ALPS									
Ortles-Cevadale Valtellina Valley									
Zebu and Chedec (Pelfini, unpublished)									
S 945/75/995–1190					>———————				
S 940/70/1100–1185						>———			
Genova Valley (Aeschlimann, 1983)									
(GX-15138)1190/75/725–995			>============						
MONT BLANC GROUP									
Miage									
S(UZ-296)760/50/1220–1265								—·<	
S(UZ-334)690/60/1250–1350								———<	
	600	700	800	900	1000	1100	1200	1300	1400AD

Note: W = wood, S = soil, M = moss, P = peat, === = in situ.

ination of earlier work by the Zürich group is under way and may be expected to amplify the present picture.

Evidence from other parts of the Alps is much more fragmentary, and absolute dates from trees *in situ* are not, so far, available to reduce the degree of uncertainty involved in the individual radiocarbon results. It is obvious from Table I that less data is forthcoming from the rest of the Alps than from the Aletsch and the other three glaciers discussed already, quite apart from that collected by Röthlisberger from other Swiss glaciers. The fluctuations of Austrian, Italian and French glacier tongues have been in phase with each other since deliberate monitoring began.

There is as yet no direct evidence from the Austrian Alps of the MWP advances in the 12th or 13th centuries, though this is likely to be forthcoming as investigations continue. In the Italian Alps, well-developed soils from under moraines marking the maximum extent of the Chedec and Zebu glaciers, in the Ortles-Cevadale group, have been dated to 945±75 BP and 940±70 BP (Pelfini, unpublished). This indicates that the two moraines were deposited after 995–1190 A.D. and after 1100–1185 A.D. at 68% probability (Table I). Only the top centimetre was dated in each case. In the Genova valley, the top of a layer of peat underlying a Holocene moraine provides a maximum date for moraine formation of 1190±75 BP (GX-13518), or 725–995 A.D. In the Mount Blanc group a pre-LIA advance followed by recession is indicated by two dates on buried soils of 760±50 BP (UZ-396) and 690±60 BP (UZ-334), that is 1220–1265 A.D. and 1250–1350 A.D., at 68% probability. A buried soil in the forefield of the Trient (in the Swiss sector of the

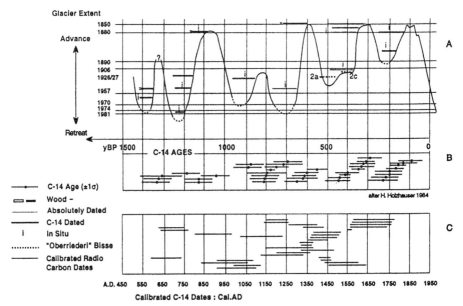

Fig. 3. The oscillations of the Grosser Aletsch Glacier, Switzerland, since 2000 years BP: (A) Oscillations of the glacier since 1500 BP, shown in comparison with positions occupied by the ice since 1850 AD. The graph was based on a variety of evidence, including the positions of absolutely dated trees *in situ*, as well as radiocarbon dated samples, many of them also *in situ*. The great majority were found in lateral positions above the present ice surface. The height above the glacier of the positions where the samples were found formed an essential element of the evidence used in the reconstruction. Most of them were taken from the outer rings of trees which were overrun by the ice. Very full details are recorded in Holzhauser (1984); (B) Conventional radiocarbon ages (BP) as presented in Holzhauser (1984); (C) Calibrated date-ranges (cal. AD) for key samples from the Grosser Aletsch (listed in Table I). During the past few centuries the radiocarbon content of the atmosphere has varied rather widely so that the measurement of the activity of organic samples growing during this period is not a good guide for their age. Even high precision measurements leading to calculations of radiocarbon ages with small uncertainties of the order of 20 years when calibrated produce a wide range of possible calendar dates extending to AD 1950, the zero year of the radiocarbon time scale. Such samples should be reported as *modern*. For this reason samples with conventional radiocarbon ages of ca 200 years or younger have been omitted from the calibrated section of Figure 2.

Mount Blanc group) dates at 825+50 BP or 1065–1260 A.D. (UZ-233). None of this evidence is in conflict with better controlled Swiss evidence.

Documentary evidence also provides some support. At the Rutor glacier, floods have occurred when the glacier was in extended positions and formed a dam. The earliest floods seem to have taken place in 1371–1470, and even perhaps as early as 1284 A.D. (Sacco, 1917). Floods definitely occurred in all years around 1594–98 A.D. and later during the LIA proper. The fourteenth century event thus signals that the ice was similarly advanced and that here, at least, the MWP was over. It has also been suggested that it was the bursting of the Lac de Combal, dammed by the extension of the Miage glacier, which caused floods before 1340 A.D. (Rabot, 1902). No indications are available from Rutor of the time when the MWP began.

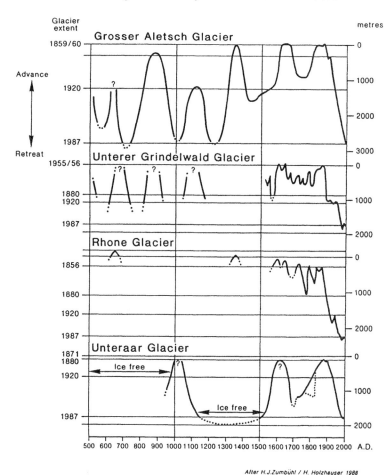

Fig. 4. Oscillations of other Swiss glaciers compared with those of the Grosser Aletsch Glacier. (From Holzhauser, 1988).

The most closely dated glacial geological evidence for the timing of the MWP comes from Switzerland. This indicates that the MWP was preceeded by an advance phase culminating between about 850–900 A.D., which brought fronts forward to positions comparable to those of the LIA maxima, that it was interrupted by a brief period of limited glacial expansion, culminating between about 1100–1150 A.D., and that it was terminated by a climatic deterioration which caused fronts to reoccupy advanced positions by about 1350 A.D. On the basis of current knowledge, a European Science Foundation working group on glacier fluctuations recently concluded that the MWP occurred between about 900 A.D. and 1250–1300 A.D. in the European Alps (Grove and Orombelli, in press).

TABLE II: Key dates relating to the glacier oscillations in Norway and the Himalaya. The numbers at the ma█ give the laboratory identification number, the C14 age and the calibrated age for 68% probability. Horizontal l█ represent calibrated dates. The symbol > indicates a maximum date; < indicates a minimum date

	600	700	800	900	1000	1100	1200	1300	1400A█
NORWAY									
Hardangerjoklen									
Hlaisen (Nesje and Dahl, 1991)									
P(T-9203)1130/70/810–985			>=======						
P(T-8204)1040/60/950–1025					—<				
Jostedalsbreen									
Sandskardfonna Cirque Glacier, Sunndal (Nesje *et al.*, 1991)									
W(T-8528)1270/60/660–860	xxxxxxxxxxxx								
W(T-5814)1180/70/755–960		xxxxxxxxxx							
W(T-8517)1110/60/880–990			xxxxxxx						
W(T-5813)890/60/1025–1225						xxxxxxxxxxxxx			
W(T-10028)620/70/1280–1405								>xxxxxxxx	
Jotunheimen									
Storbreen (Griffey and Matthews, 1978)									
M(SRR-1083)644/45/1255–1375							>———		
M(SRR-1084)532/40/1380–1435								>——	
Svartisen									
Engabreen (Worsley and Alexander, 1975) – *in situ* trees									
W(HAR-385)1060/80/880–1025			>=======						
W(HAR-386)1230/80/660–890	>==========								
Fingerbreen									
P(I-10364)695/75/1245–1375							>——-		
HIMALAYA									
Khumbu (Benedict, 1976)									
S(Hv-12009)960/70/990–1155					>———				
Khumbu (Röthlisberger, 1986)									
(B-174)480/80/1385–1465								—<	
Lhotse Nup (Röthlisberger, 1986)									
S(Hv-12004)1025/110/880–1110				>———					
S(Hv-10738)1190/155/650–1010	—————<								
Lhotse Shar									
S(Hv-12006)775/110/1050–1285						>———			
S(Hv-10743)1205/50/750–885		——<							
S(Hv-12005)610/40/1275–1390								>——	
S(Hv-10742)1135/70/805–980			————<						
Nupse									
(Hv-10746)565/55/1295–1420								————<	
Gilgit Region – Gargo									
(Hv-10723)585/80/1265–1430								————<	
	600	700	800	900	1000	1100	1200	1300	1400AI

Note: W = wood, S = soil, M = moss, P = peat, === = in situ, xxxx = warm period.

7. Scandinavia

If the evidence from the Alps is in general agreement, it is logical to consider next whether the same is true of data from Scandinavia (Table II). Current data is very widely scattered and is not abundant.

Tree stumps have been found in paleosols interlayered with distal glaciofluvial fan deposits at Sandsvora, Sunndalen, at the northern end of Jostedalsbreen. Radiocarbon dates from these organic layers indicate that the Sandskardfonna cirque glacier had either contracted or had disappeared close to 1270±60 BP (T-8528), 1180±70 (T.5814), 1110±60 (T-8527), 890±60 (T-5813) and 620±70 (1280–1405) (Table II). It seems that the glaciation threshold must have been at least 50 m above its present altitude during intervals when Sandskardfonna was melted away. The youngest of the dates obtained gives a maximum date of 1280–1405 A.D. for the first LIA expansion of the Sandskardfonna cirque glacier (Nesje *et al.*, 1991; Nesje and Rye, 1993).

The forefield of Blaisen, a north-eastern outlet of Hardangerjoklen, (69°33'N, 7°25'E) has been studied by Nesje and Dahl (1991). A stratigraphic sequence of bluish-grey sand and silt, taken to be of glaciofluvial origin, is interbedded with organic material, deposited during intervals of frontal recession. A thin sand/silt deposit near the top of the studied section lies between peat layers. A sample from the upper part of the underlying peat gives a date of 1130± 70 BP (T-9203) and a maximum age for a short glacial advance preceeding the MWP of 810–985 A.D. A sample from the overlying peat of 1040±60 BP (T-8204) provides a minimum date for deposition of the fluvioglacial sediments, which may have occurred at about the same time as the advance of the Aletsch in the mid-MWP, and a maximum but not necessarily close date for later Little Ice Age advances of 950–1025 A.D. Other evidence from southern Scandinavia, coming from the Jotunheimen (Griffey and Matthews, 1978), provides closer dating for advances after the MWP. Moss from underneath a LIA moraine of Storbreen has dates of 664±45 BP (SRR-1083), and 532±40 BP (SRR-1084), giving maximum ages of 1255–1375 A.D. and 1380–1435 A.D. respectively, fitting well with the results from Sandskardfonna.

While no unequivocal documentary evidence of the MWP has been found in southern Norway, it has been suggested that severe damage to farms in Olden and Loen parishes, situated immediately to the north of Jostedalsbreen in the first half of the fourteenth century, may well have been associated with glacial expansion and associated increases in landslides and floods (Grove, 1985).

A sample from the top 2 cm of peat under a moraine fronting Fingerbreen, an eastern outlet of Ostisen, in the Svartisen region of northern Norway, of 695±75 BP (I-10364) indicates that deposition took place after 1245–1375 A.D. (Karlén, 1988). The closest dates available from Norway at present are those from *in situ* alder of 1060±80 BP (HAR-385) and a willow stump, also *in situ*, 1230+/080 BP (HAR-386), both near Engerbreen (Worsley and Alexander, 1975). These dates of

880–1025 A.D. and 660–890 A.D. for glacial advance preceeding the MWP are not in conflict with the Alpine reconstruction.

Although the Scandinavian evidence is both fragmentary and scattered most of that which is available so far suggests that glacial history during the last millennium was probably roughly similar to that in the Alps. But it must be noticed that the Sandskardfonna evidence is interpreted by Nesje and Rye as showing that while there were small bursts of glacial activity alternating with warmer periods of little or no such activity from around 660 BP onwards until after 1280–1405 A.D. when the first LIA advance began, no clear evidence of major glacial expansion before the MWP was forthcoming. This suggests the possibility of a real difference in conditions between southern Norway and the north of the country, where the Fingerbreen data seems to fit with that from Switzerland. More investigations are needed if a more definitive view is to be obtained.

8. The Himalaya

Dating of LIA initiation in the Mt. Everest region (Table II) depends on studies of the moraines of only a very few glaciers. The stratigraphic relationships of some of the radiocarbon dated samples which have been collected are somewhat uncertain (Williams, 1983), and have therefore not been included. Röthlisberger (1986) examined the moraines of 16 glaciers in the Himalaya, selecting only those comparable in size with those in the Alps. Dates from paleosols amongst the moraines of three glaciers in the Mt. Everest area and one in the Gilgit region are shown on Table II (Benedict, 1976; Röthlisberger and Geyh, 1985a; Röthlisberger 1986). These provides evidence of moraine formation after 1050–1285 A.D. and before 1110 A.D. The moraine formed after 1050–1285 A.D. could have been contemporary with the minor advance during the MWP in the Alps, but could also belong to the first LIA advances, which were evidently under way between 1265–1465 A.D. These few dates, all on soil, are inadequate guides to the timing of glacial events in the Himalaya, which contains the largest ice-covered area outside the Polar regions, but do not in themselves indicate any significant difference from those in the European Alps.

9. North America

9.1. *Alaska*

The best known glaciers in this enormous region are either surging or tidewater glaciers and so dating their fluctuations is not a reliable guide to the timing of the MWP. Röthlisberger investigated moraines on a number of forefields and added considerably to knowledge of land-based glaciers (Röthlisberger and Geyh, 1986). Dates from several moraine sets around the Juneau Icefield are shown on Table III. *In situ* stumps on the side of a nunatak of the Llewellyn Glacier indicate kill

TABLE III: Key dates relating to the glacier oscillations in North America. The numbers at the margin give the laboratory identification number, the C14 age and the calibrated age for 68% probability. Horizontal lines represent calibrated dates. The symbol > indicates a maximum date; < indicates a minimum date

	600	700	800	900	1000	1100	1200	1300	1400AD
ALASKA AND YUKON									
Juneau Icefield (Röthlisberger and Geyh, 1986)									
Gilkey									
W(Hv-12095)1195/129/655–990	>————————								
S(Hv-11300)1000/85/955–1095					>———				
S(Hv-11991)760/85/1165-1275							>———		
W(Hv-11299)930/85/995–1220					>=============				
W(Hv-11301)565/55/1295–1420								>=========	
W(Hv-12094)515/50/1385–1445									>—
Hoboe									
S(Hv-11318)1060/75/880–1025				>———					
W(Hv-11312)1025/60/955–1030				<?>xxx					
W(Hv-11319)430/55/1425–1470									?xxx
S(Hv-11317)425/55/1430–1480									>—
Llewellen									
W(Hv-11365)690/80/1245-1380								>======	
W(Hv-11364)465/50/1405–1465									?xxx?
W(Hv-11369)355/70/1450–1645									>=====
CANADIAN CORDILLERA									
Coast Mountains (Ryder and Thomson, 1986)									
Franklin									
W(S-1568)835/45/1070–1250						>———			
Bridge									
W(S-1463)680/50/1250–1350								>———	
W(S-1571)540/45/1375–1435									===
Sphinx									
W(Y-347)476/60/1395–1460									====
Scud (Ryder-quoted Osborn and Luckman, 1988)									
W(S-1298)625/40/1265–1385								>—	
W(S-2297)455/65/1400–1465									>—
Interior Ranges									
Bugaboo (Osborn, 1986)									
W(B-7576)900/60/1020–1220					>———				
Big Eddy Creek (Wheeler, 1964)									
W(GSC-571)450/130/1370–1635									>——
Niwa (Luckman, 1986)									
W(B-12416)840/80/1035–1260					>===========				
	600	700	800	900	1000	1100	1200	1300	1400AD

Note: W = wood, S = soil, M = moss, P = peat, === = in situ, xxxx = warm period.

TABLE III: *Continued*

	600	700	800	900	1000	1100	1200	1300	1400AD
Rocky Mountains									
Robson (Osborn and Luckman, 1988)									
W(B-9458)1140/80/795–990			>=========						
in situ tree, +373 for tree ring age									
W(B-3731)900/60/1020–1220						>==========			
in situ tree, +40 for tree ring age									
W(B-2816)860/50/1040–1230						>=========			
in situ tree, +115 for tree ring age									
	600	700	800	900	1000	1100	1200	1300	1400AD

Note: W = wood, S = soil, M = moss, P = peat, === = in situ, xxxx = warm period.

by advances of the ice at the end of the MWP in 1245–1380 A.D. and again in 1450–1645 A.D. The original position of the wood fragment which was growing in 1405–1465 A.D. is unknown.

The dating of the top of a fossil soil in the right lateral moraine of the Hoboe (Hv-11318) provides a maximum age for the formation of the moraine above it by an advance preceeding the MWP between 880 A.D. and 1025 A.D., while warm periods between 955–1030 A.D. and 1425–1470 A.D. may be indicated by samples of wood not *in situ* (Hv-11312 and Hv-11319). Another soil sample from the right lateral moraine (Hv-11317) seems to yield evidence of moraine formation around or after 1430–1480 A.D., after the MWP. None of these dates can be taken as close however. The evidence from the Gilkey is the most complete for the period of interest. The samples Hv-11299 and Hv-11301 both come from trees *in situ*, providing clearer evidence of the time of kill, but neither the number of rings displayed by these stumps nor the positions of the samples in relation to the rings were published.

9.2. *Canadian Cordillera*

The available evidence for the timing of the MWP in the Canadian Cordillera is also shown on Table III. On the basis of this assemblage of dates, Osborn and Luckman (1988) proposed that an advance period, which they termed the Cavell, occurred in the region shortly after 900 BP An example of the type of evidence is the date for a moraine of the Bugaboo Glacier of 900±60 BP (B-7576) obtained by Osborn (1986) from a tree limb, indicating an advance after 1020–1220 A.D.

Possible evidence of an earlier phase of expansion comes from Robson Glacier, where Hauser (1956) discovered a fossil spruce forest which had been felled by the ice and now lies beneath three metres of till (Luckman, 1986). Several of the trees have been sampled; the oldest (B-9458) has a calendar age of 795–990 A.D., but as it was 373 years old, the time it was overrun was in the period 1168–1363 A.D. (assuming that the innermost part was that sampled by ^{14}C). Cross-dating of the rings may in due course reveal whether the glacier advance which killed it was coincident with those in the Alps or earlier. The fossil forest was growing in a thin paleosol above a fine-grained grey till. These units could represent soil formation during the MWP and a preceding glacial expansion. A similar picture comes from investigation of the lateral moraines of the Klinakini and Franklin Glaciers in the Coast Mountains (Ryder and Thomson, 1986), but no clear indications are available about the date of initiation of the paleosols.

Moraines formed in the later phases of the LIA are commonly so large in Canada as to conceal older and smaller deposits (Luckman, 1993). In Canada, as in Switzerland, dating of trees *in situ* is providing a more accurate approach to dating than would be possible on the basis of radiocarbon alone (Luckman, 1986). As many more stumps *in situ* are being uncovered by the current rapid recession, it may well become possible to reconstruct the timing of glacier fluctuations in the Canadian Cordillera more completely than can be done at present. (Luckman, 1993).

Evidence relating to the MWP has been found in widely scattered locations in North America. This suggests that the timing of the glacial advances preceeding and closing it may well have been in general accordance with that in the European Alps, but it is not yet sufficiently dense for it to be possible to identify regional variations satisfactorily.

10. Tropical South America

Well-dated evidence of the positions and extents of the tropical glaciers during and since the medieval centuries is sparse (Table IV), in part because many of them are small and in very remote places, difficult of access.

An advance of the Quelccaya Icecap, Cordillera Occidental, Peru, around 905±100 BP (I-8441) or 1000–1245 A.D., has been recognised on the basis of an exposure of peat bulldozed by ice readvance (Mercer and Palacios, 1977). Data from three glaciers in the Cordillera Blanca, Peru, was collected from exposures in lateral moraines by Röthlisberger and Geyh (1986). The oldest samples, from Huallacacocha, were of wood from trees *in situ*, which were killed sometime around or after 625–795 A.D. and 730–880 A.D. A further sample from a paleosol within the same moraine (Hv-8703) suggests the occurence of a later advance, closing the MWP, sometime after 1255–1335 A.D. The paleosol was 50–65 cm thick, indicating a long period of soil formation, with no disturbance from the ice.

Another paleosol sample from a lateral moraine of the Ocshapalca, with a date of 440±185 BP, gives a maximum age for the moraine above but is insufficiently precise to be very useful in supporting the Huallacacocha data. The single date of 500±80 BP on peat from the Hichu-Khota valley, Cordillera Real, Bolivia, indicates moraine formation after 1375–1460 A.D. but does not preclude the possibility

TABLE IV: Key dates relating to the glacier oscillations in South America. The numbers at the margin give the laboratory identification number, the C14 age and the calibrated age for 68% probability. Horizontal lines represent calibrated dates. The symbol > indicates a maximum date; < indicates a minimum date

	600	700	800	900	1000	1100	1200	1300	1400AD
TROPICAL SOUTH AMERICA									
PERU (Wright, 1984; Gouze *et al.*, 1986; Röthlisberger and Geyh, 1986; Seltzer, 1990									
Mercer and Palacios, 1977)									
Quelcaya Icecap									
P(I-8441)905/100/1000–1245					>———————				
Cordillera Blanca									
Huallacacocha									
W(Hv-8710)1325/25/625–795	>=======								
W(Hv-8709)1215/45/730–880		>======							
S(Hv-8703)690/50/1255–1335								>=======	
Ocshapalca									
S(Hv-8704)440/185/1270–1660								>———————	
Cuchapanga									
P(WIS-1031)1100/70/875–1005				>———					
P(WIS-1200)430/70/1410–1525									>———
Laguna Huatacocha									
P(WIS-1031)1100/70/875–1005				>——					
Nevado Huaytapallana									
P(SI-6996)535/105/1270–1460								>———	
BOLIVIA									
Cordillera Real									
Pichu-Khota Valley									
H(I-?)500/80/1375–1460									>———
EXTRATROPICAL SOUTH AMERICA									
CHILE									
Cordillera Central – Cipreses									
W(Hv-10915)625/155/1225–1450							>———		
ARGENTINA									
Tronador – Ventisquero Negro del Rio Manso									
W(Hv-11800)940/110/965–1225					>===============				
W(Hv-12865)620/50/1265–1390								>======	
W(Hv-12864)585/50/1280–1405								>======	
W(Hv-11799)300/85/1460–1670									>======
	600	700	800	900	1000	1100	1200	1300	1400AD

TABLE IV: *Continued*

	600	700	800	900	1000	1100	1200	1300	1400AD
CHILE									
Lago O'Higgins									
Ventisquero Huemal									
W(Hv-10898)1345/105/600–805	>———								
W(Hv-10897)1325/55/645–760	>——								
W(Hv-10896)1285/55/655–805	>———								
W(Hv-10893)895/85/1010–1230					xxxxxxxxxxxxx				
W(Hv-10892)470/55/1400–1460									xxxxxxx
W(Hv-10894)465/65/1395–1465									>——
W(Hv-10889)425/45/1435–1470									xxxxxx
Ventisquero Bravo									
W(Hv-10899)665/80/1255–1390								>——	
Cordillera del Paine									
Ventisquero Perro									
W(Hv-10887)795/80/1075–1265						>==========			
W(Hv-10886)345/100/1440–1660									>=======
	600	700	800	900	1000	1100	1200	1300	1400AD

Note: W = wood, S = soil, M = moss, P = peat, === = in situ, xxxx = warm period.

that the LIA advances may have begun earlier. A certain amount of information may be obtained from investigation of the timing of trim-line formation. In the Cerros Cuchpanga, Peru, a clear trim-line about 100 m below the present ice-front separates bare rock and till from puna grassland. Ages of 1100±70 BP (WIS-1031) and 430±70 BP (WIS-1200) from the base of peat formed on outwash fans of two small glaciers have been interpreted as providing minimum dates for the same phase of glacial recession (Wright, 1984). The difference in age is attributed to outwash deposition continuing longer in the latter case. If this interpretation is correct, it serves to demonstrate very well the need for detailed field observations if false conclusions are to be avoided. Seltzer (1990) argued that dates from the base of peats near a trim-line at Nevado Huaytapallana (Si-6996) could also imply a phase of glaciation starting before 1300 A.D.

11. Extra-Tropical South America

Mercer (1965, 1968, 1970, 1976) was responsible for extensive field studies in South America. Further invesigations, concentrating particularly on lateral moraine exposures, were made by Röthlisberger (1985b, 1986). All the data on Table IV comes from Röthlisberger (1986). The dates selected for discussion are those with the clearest implications.

In the moraines of the Ventisquero Negro or Rio Manso, on the south-east side of Cerro Tronador (42° S) trunks of *Nothofagus* remain *in situ*. Röthlisberger concluded that four dates from the trees revealed in a river cut through the left lateral moraine provide evidence of several advances. The date of 1265–1390 A.D. (Hv-10915) from the Glaciar Cipreses, some 8° latitude further north in the Cordillera Central, is not inconsistent with those from the Rio Manso, but came from wood which was not *in situ*.

Wood, *in situ*, from a fossil forest in paleosol 60–70 cm thick on bedrock, from a section through a lateral of the Ventisquero Bravo, in the O'Higgins region of Chile (40° S) has an age of 665±80 BP (Hv-10899). Five more dates from Ventisquero Huemal, in the same region, are all from wood not *in situ*. Hv-10897 and Hv-10896 come from wood in a fossil soil 100 m below the crest of a moraine and 100 m above the ice surface. The individual trunks were up to 40–50 cm in diameter. Each was taken to represent a glacial advance around or after the indicated ages. Röthlisberger considered that the date of 895±85 BP (Hv-10893) and the three younger ones indicate times of minimum extension of the ice. All the wood samples were found some 300 m in front of the present glacier, its original position unknown, so the implications are not firmly based.

The final data available are from the Cordillera del Paine, Chile, at 51° S. Both dates are from *in situ* trunks overridden by ice. These two different ages from a sheltered site were considered by Röthlisberger to represent separate advances.

Not one of these data sets includes a pair of bracketing dates for the same event, but the very different times at which trees were overridden by ice clearly shows that several advance phases occurred. These must have been separated by intervals during which the climate was suitable for tree growth. The general accordance of the results of investigations in very widely separated regions within South America is striking. There is no doubt that much more information could be obtained from further fieldwork among the many small sensitive glaciers in the Andes.

12. New Zealand

New Zealand data are shown on Table V. The glaciers concerned are all in South Island, either in Westland or in the much drier region to the east of the New Zealand Alps, in the Mount Cook or Rangitata areas.

Wood and fossil soils embedded in the superposed lateral moraines of the Tasman Glacier have yielded a considerable amount of information. On the basis of this a direct comparison has been made between the timing of the fluctuations of the Tasman front and that of the Aletsch during the last 1500 years (Gellatly *et al.*,1985). Interpretation of the evidence is not entirely straightforward because of the complexity of the moraine sections involved. The comparison with the Aletsch is persuasive but requires further support.

All the evidence from the Balfour Glacier comes from stumps *in situ*. A sample (Hv-10523), dated at 550–655 A.D., came from 10 m below the crest of a lateral

TABLE V: Key dates relating to the glacier oscillations in New Zealand. The numbers at the margin give the laboratory identification number, the C14 age and the calibrated age for 68% probability. Horizontal lines represent calibrated dates. The symbol > indicates a maximum date; < indicates a minimum date

	600	700	800	900	1000	1100	1200	1300	1400AD
NEW ZEALAND									
Mount Cook									
Tasman (Burrows, 1973, 1979; Röthlisberger and Geyh, 1986; Gellatly *et al.*, 1988)									
S(Hv-10490)1075/40/885–1000				>—					
W(NZ-4509)970/60/990–1100					—<				
S(Hv-12018)900/60/1020–1220					>———				
S(NZ-5331)864/50/1040–1225					>———				
S(Hv-10500)765/50/1215–1265							>—		
S(NZ-711)684/48/1255–1340							>——		
S(Hv-12027)550/70/1290–1440								>————	
W(Hv-711)520/60/1375–1450									>——
S(Hv-12024)525/55/1375–1445									>——
S(Hv-10500)765/50/1215–1265						—<			
W(NZ-5330)343/50/1455–1640									>—
Mueller (Röthlisberger and Geyh, 1986; Burrows, 1980)									
S(NZ-4507)1010/50/965–1030					——				
Westland									
Horace Walker (Wardle, 1973; Röthlisberger and Geyh, 1986)									
P(NZ-2329)970/50/995–1065					>——				
P(NZ-3929)940/50/1010–1150					———				
Balfour (Burrows, 1979; Burrows and Greenland, 1979)									
W(Hv-10523)1440/55/550–655									
W(Hv-10522)1260/55/660–875		>==========							
W(Hv-10521)905/70/1015–1220					>===========				
W(Hv-10524)720/45/1250–1270							>==		
W(Hv-11278)530/55/1375–1445									>====
Rangitata River									
McCoy									
W(NZ-4774)664/77/1255–1385							>——		
Colin Campbell									
W(NZ-4105)650/60/1255–1385							>——		
W(NZ-4016)520/60/1375–1450								>—	
	600	700	800	900	1000	1100	1200	1300	1400AD

Note: W = wood, S = soil, M = moss, P = peat, === = in situ, xxxx = warm period.

moraine; Hv-10522, dated at 660–875 A.D., came from 100 m below the same crest. A sample (Hv-10521) was from a stump 15 m below the crest, and the two youngest were found 40 m and 50 m below it. It is reasonable to conclude that each one reflects an episode of shearing by the ice soon after the dates given by the samples, the lag depending on the unknown ages of the trees when they were killed. More clement conditions evidently intervened between the episodes of glacial expansion, during which trees could grow. It is unfortunate that Hv-10521 gives such a wide range of calendar dates as to make it impossible to determine whether it relates to an advance during the MWP or at the beginning of the LIA.

The accordance of the New Zealand data with the Swiss is rather clearer than that from other regions, although there is plenty of room for improvement and tighter control.

13. Conclusion

The first phase of the LIA began around the thirteenth century in all the regions for which there is evidence. The glacial phase preceding the MWP seems to have begun between the seventh and ninth centuries A.D. but is generally less securely dated and not dated at all in Canada. There are at least some indications of fluctuations in ice position in the course of the MWP in Norway, Alaska, and perhaps in extratropical South America and New Zealand, indicating that recession may have been interrupted by advances, perhaps of limited extent, as in the European Alps. The available evidence suggests that the MWP was global in extent and not uniform climatically. The glacial data needs to be considered in relation to that from other sources, but is of value in obtaining a more complete understanding of both the environment in the later medieval period and the possible causes of climatic change on the century time scale.

Acknowledgements

We wish to thank Jennifer Wyatt and Ian Agnew of the Department of Geography, Cambridge University and Simon Crowhurst of the Godwin Laboratory for drawing the diagrams.

References

Aeschlimann, H.: 1983, *Zur Geschichte des Italienischen Mont Blanc Gebietes: Val Veni-Val Ferret-Ruitor*, Unpublished Dissertation, Geographischen Institut der Universität Zürich.
Aitchison, T. C. *et al.*: 1989, 'A Comparison of Methods Used for the Calibration of Radiocarbon Ages', *Radiocarbon* **31**, 846–64.
Benedict, J. B.: 1976, 'Khumbu Glacier Series', *Radiocarbon* **18**, 177–8.
Bless, R.: 1984, 'Beitrage zur Spät- und Postglazialen Geschichte der Gletscher im Nordöstlichen Mont Blanc Gebiet', *Phys. Geogr.* **15**, 1–116.
Bradley, R. S. and Jones, P. D.: 1992a, 'When Was the "Little Ice Age"', in Mikami, T. (Ed.), *Proceedings of the International Symposium on the Little Ice Age Climate*, Tokyo Metropolitan University.
Bradley, R. S. and Jones, P. D.: 1992b, *Climate since A.D. 1500*, Routledge, London and New York.
Bradley, R. S. and Jones, P. D.: 1993, ' "Little Ice Age" Summer Temperature Variations: Their Nature and Relevance to Recent Global Warming Trends', *The Holocene* **3**, 367–76.
Burrows, C. J.: 1973, 'Studies of Some Glacial Moraines in New Zealand – 2. Ages of Moraines of the Mueller, Hooke and Tasman Glaciers (S79), *New Zeal. Geol. Geophys.* **16**, 831–55.

Burrows, C. J.: 1979, 'A Chronology for Cool-Climate Episodes in the Southern Hemisphere 12000–1000 yr. BP', *Palaeogeogr. Palaeoclimatol. Palaeoecol.* **27**, 287–347.

Burrows, C. J.: 1980, 'Radiocarbon Dates for Post-Otiran Glacial Activity in the Mt. Cook Region, New Zealand', *New Zeal. J. Geol. Geophys.* **23**, 239–48.

Burrows, C. J. and Greenland, D. E.: 1979, 'An Analysis of the Evidence for Climatic Change in New Zealand in the Last Thousand Years: Evidence from Diverse Natural Phenomena and from Instrumental Records', *J. Roy. Soc. New Zeal.* **9**, 321–73.

Calkin, P. E.: 1988, 'Holocene Glaciation of Alaska (and Adjoining Yukon Territory, Canada)', *Quatern. Sci. Rev.* **7**, 159–84.

Calkin, P. E. and Ellis, J. M.: 1981, 'A Cirque Glacier Chronology Based on Emergent Lichens and Mosses', *J. Glaciol.* **27**, 512–5.

Clapperton, C. M. and Sugden, D. E.: 1988, 'Holocene Glacier Fluctuations in South America and Antarctica', *Quatern. Sci. Rev.* **7**, 185–98.

Egan, C. P.: 1971, *Contribution to the Late Neoglacial History of the Lynn Canal and Taku Valley Sections of the Alaska Boundary Range*, PhD Dissertation, State University, East Lancing, Michigan.

Fushini, H.; 1977, 'Glaciation in the Khumbu Himal', *Seppyo, J. Jap. Soc. Snow Ice* **39**, 60–7.

Fushini, H.: 1978, 'Glaciation in the Khumbu Himal', *Seppyo, J. Jap. Soc. Snow Ice* **40**, 71–7.

Gellatly, A. F., Chinn, T. J. H., and Röthlisberger, F.: 1988, 'Holocene Glacier Variations in New Zealand: A Review', *Quatern. Sci. Rev.* **7**, 227–42.

Griffey, N. J. and Matthews, J. A.: 1978, 'Major Neoglacial Expansion Episodes in Southern Norway: Evidence from Moraine Stratigraphy with C14 Dates on Buried Paleosols and Moss Layers', *Geograf. Annal.* **60**(A), 73–96.

Grove, J. M.: 1985, 'The Timing of the Little Ice Age in Scandinavia', in Tooley, M. J. and Sheail, G. M. (eds.), *The Climatic Scene*, Allen and Unwin, London.

Grove, J. M.: 1988, *The Little Ice Age*, Methuen, London and New York.

Grove, J. M. and Orombelli, G.: 'The Southern Region of Europe; the Alps, Italy and the Pyrenees', in Boulton, G. S. and Mason, P. (eds.), *Evaluation of Climate Proxy Data in Relation to the European Holocene*, Special issue: ESF Project European Climate and Man. *Paläoklimaforsch. Palaoclim. Res.*, in press.

Gouze, P., *et al.*: 1986, 'Interpretation Paleoclimatique des Oscillations des Glaciers au Cours des 20 Derniers Millenaires dans les Regions Tropicales: Example des Andes Boliviennes', *Compte Rend. L'Acad. Sci. Paris* **303** (II), 219–23.

Heuser, C. J.: 1956, 'Postglacial Environments in the Canadian Rocky Mountains', *Ecol. Monogr.* **26**, 253–302.

Holzhauser, P.: 1984a, 'Zur Geschichte der Aletschgletscher und des Fieschergletscher', *Phys. Geogr.* **13**, 1–448.

Holzhauser, P.: 1988, 'Methoden zur Rekonstruktion von Gletscherschwankungen', *Die Alpen* **64**, 135–65.

Holzhauser, P.: 1984b, 'Rekonstruktion von Gletscherschwankungen mit Hilfe Fossiler Holzer', *Geograph. Helv.* **39**, 3–15.

Karlén, W.: 1988, 'Scandinavian Glacial and Climatic Fluctuations during the Holocene', *Quatern. Sci. Rev.* **7**, 100–209.

Lamb, H. H.: 1965, 'The Early Medieval Warm Period and Its Sequel', *Paleogeogr., Paleoclimatol., Paleoecol.* **1**, 13–37.

Luckman, B. H. and Osborn, G. D.: 1979, 'Holocene Glacier Fluctuations in the Middle Canadian Rocky Mountains', *Quatern. Res.* **25**, 10–24.

Luckmann, B. H.: 1986, 'Reconstruction of Little Ice Age Events in the Canadian Rocky Mountains', *Geograph. Phys. Quatern.* **XL**, 17–28.

Luckman, B. H.: 1993, 'Neoglacial Glacier Fluctuations in the Canadian Rockies', *Quatern. Res.* **39**, 144–53.

Lütschg, O.: 1926, 'Über Niederschlag und Abfluss im Hochgebirge, Sonderdarstellung des Mattmarkgebietes', *Schweizerischer Wasserwirtschaftverband, Verbandschrift C14*, Veröffentlichung der Schweizerischen Meteorologischen Zentralanstalt in Zürich, Zürich Sekretariat des Schweizerischen Wasserwirtschaftstsverbandes.

Kinzl, H.: 1932, 'Die Grössten Nacheiszeitlichen Gletschervorstösse in den Schweizer Alpen und in der Mont-Blanc-Gruppe', *Zeitschr. Gletscherk.* **20**, 269–397.

Mann, D. H.: 1986, 'Reliability of a Fjord Glacier's Fluctuations for Paleoclimatic Reconstruction', *Quatern. Res.* **25**, 10–24.

Matthews, J. A.: 1984, 'Limitations of C14 Dates from Buried Soils in Reconstructing Glacier Variations and Holocene Climate', in Morner, N. A. and Karlén, W. (eds.), *Climatic Changes on a Yearly to Millennial Basis: Geological, Historical and Instrumental Records*, 281–90.

Matthews, J. A.: 1985, '14C Dating of Paleosols, Pollen Analysis and Landscape Change: Studies from Low- and Mid-Alpine Belts of Southern Norway', in Boerdman, J. (ed.), *Soils and Quaternary Landscape Evolution*, John Wiley London, pp. 87–116.

Mayr, F.: 1964, 'Untersuchungen über Ausmass und Folgen der Klima- und Gletscherswankungen seit dem Beginn der postglazialen Wärmezeit', *Zeitschr. Geomorphol.* **8**, 257–85.

Mercer, J. H.: 1965, 'Glacier Variations in Southern Patagonia', *Geograph. Rev.* **55**, 390–413.

Mercer, J. H.: 1968, 'Variations of Some Patagonian Glaciers since the Late Glacial I', *Amer. J. Sci.* **266**, 1–25.

Mercer, J. H.: 1970, 'Variations of Some Patagonian Glaciers since the Late Glacial II', *Amer. J. Sci* **269**, 91–109.

Mercer, J. H.: 1976, 'Glacial History of Southernmost South America', *Quatern. Res.* **6**, 125–66.

Mercer, J. H. and Palacios, O.: 1977, 'Radiocarbon Dating of the Last Glaciation in Peru', *Geol.* **5**, 600–4.

Miller, G. H.: 1973, 'Late Quaternary Glacial and Climatic History of Northern Cumberland Peninsula, East Baffin Island, NWT, Canada', *Quatern. Res.* **3**, 561–83.

Nesje, A.: 1991, 'Holocene Glacial and Climatic History of the Jostedalen Region, Western Norway' Evidence from Lake Sediments and Terrestial Deposits', *Quatern Sci. Rev.* **10**, 87–114.

Nesje, A. and Dahl, S. D.: 1991, 'Holocene Variations of Blaisen, Hardangerjoklen, Central South Norway', *Quatern. Res.* **35**, 25–40.

Nesje, A. and Rye, N.: 1993, 'Late Holocene Glacial Activity at Sandskard Fonna, Jostedalsbreen Area, Western Norway', *Norsk Geogr. Tidsskr.* **47**, 21–8.

Nesje, A. *et al.*: 1991, 'Holocene Glacial and Climatic History of the Jostedalsbreen Region, Western Norway; Evidence from Lake Sediments and Terrestial Deposits', *Quatern. Sci. Rev.* **10**, 87–114.

Oeschger, H. and Röthlisberger, H.: 1961, 'Datierung Eines Ehemaligen Standes des Aletschgletschers durch Radioaktivitätsmessung an Holsproben und Bemerkungen zu Holsfunden an Weiterer Gletschern', *Zeitschr. Gletscherk. Glazialgeol.* **4**, 191–205.

Orombelli, G. and Porter, S. C.: 1982, 'Late Holocene Fluctuations of Brenva Glacier', *Geogr. Fis. Dinam. Quatern.* **5**, 14–37.

Osborn, G.: 1986, 'Lateral Moraine Stratigraphy and Neoglacial History of Bugaboo Glacier, British Columbia', *Quatern. Res.* **26**, 171–8.

Osborn, G. and Luckman, B. H.: 1988, 'Holocene Glacier Fluctuations in the Canadian Cordillera (Alberta and British Columbia)', *Quatern. Sci. Rev.* **7**, 115–28.

Porter, S. C.: 1981, 'Lichenometric Studies in the Cascade Range of Washington: Establishment of *Rhizocarpon Geographicum* Growth Curves on Mt. Rainier', *Arc. Alp. Res.* **13**, 11–23.

Porter, S. C. and Orombelli, G.: 1986, 'Glacier Contraction during the Middle Holocene in the Western Italian Alps: Evidence and Implications', *Geol.* **13**, 296–8.

Rabot, C.: 1902, 'Essai de Chronologie des Variations Glaciaires', *Bull. Geogr. Histor. Descript.* 285–327.

Röthlisberger, F.: 1986, *10000 Jahre Gletschergeschichte der Erde*, Mit einem Beitrag von M. A. Geyh, Sauerländer, Aarau und Frankfurt am Main.

Röthlisberger, F. and Schneebeli, W.: 1979, 'Genesis of Lateral Moraine Complexes, Demonstrated by Fossil Soils and Trunks: Indicators of Postglacial Climatic Fluctuations', in Schluchter, E. (ed.), *Moraines and Varves*, Balkema, Rotterdam, pp. 387–419.

Röthlisberger, F., *et al.*: 1980, 'Holocene climatic Fluctuations. Radiocarbon Dating of Fossil Soils (fAh) and Wood from Moraines and Glaciers in the Alps', *Geogr. Helv.* **35**, 21–52.

Röthlisberger, F. and Geyh, M. A.: 1985a, 'Glacier Variations in the Himalayas and Karakorum', *Zeitschr. Gletscher. Glazialgeol.* **21**, 237–49.

Röthlisberger, F. and Geyh, M. A.: 1985b, 'Gletscherschwankungen der Nacheiszeit in der Cordillera

Blanca (Peru) und den Südlichen Anden Chiles und Argentiniens', *Zentralbl. Geol. Paleoontol.* **11**, 1611–3.

Ryder, J. M. and Thomson, B.: 1986, 'Neoglaciation in the Southern Coast Mountains of British Columbia: Chronology of Events Prior to the Late Neoglacial Maximum', *Canad. J. Earth Sci.* **23**, 273–87.

Sacco, F.: 1917, 'I Ghiacciai Italiani del Gruppo Monte Bianco', *Boll. Comit. Glaciol. Ital. Ser. 2.* **3**, 21–102.

Seltzer, G. O.: 1990, 'Recent Glacial History and Paleoclimate of the Peruvian-Bolivian Andes', *Quatern. Sci. Rev.* **9**, 137–52.

Stuiver, M. and Pearson, G. W.: 1986, 'High Precision Calibration of the Radiocarbon Timescale A.D. 1950–500 B.C.', *Radiocarbon* **28**, 805–38.

Stuiver, M. and Pearson, G. W.: 1993, 'High-Precision Bidecadal Calibration of the Radiocarbon Timescale A.D. 1950–500 B.C.', *Radiocarbon* **35**, 1–23.

Suess, H. E.: 1970, 'Bristlecone Pine Calibration of the Radiocarbon Timescale 5200 B.C. to Present', in Olson, I. U. (ed.), *Radiocarbon Variations and Absolute Chronology*, Interscience, New York.

Tuthill, S. J., Field, W. O., and Clayton, L.: 1966, 'Past Earthquake Studies of the Sherman and Sheridan Glaciers', in *The Great Alaskan Earthquakes of 1964* (Hydrology Volume), National Academy of Sciences, Washington, D. C. Publication 163, pp. 318–28.

Villalba, R.: 1990, 'Climatic Fluctuations in Northern Patagonia during the Last Thousand Years as Inferred from Tree Rings', *Quatern Res.* **34**, 346–60.

Wardle, P,: 1973, 'Variations of Glaciers in Westland National Park and the Hooker Range', *New Zeal. J. Botany* **11**, 349–88.

Wheeler, J. O.: 1964, 'Selkirk and Monash Mountains: Recent Glacier Fluctuations', *Canad. Geophys. Bull.* **17**, 1267.

Williams, V. S.: 1983, 'Present and Former Equilibrium Line Altitudes near Mt. Everest, Nepal and Tibet', *Arc. Alp. Res.* **15**, 201–11.

Willis, E. H., Munnich, K. O., and Tauber, H.: 1960, 'The Variations in Atmospheric Radiocarbon Activity over the Past 1800 Years', *Amer. J. Sci. Radiocarb. Suppl.* **2**, 1–4.

Worsley, P. and Alexander, M.: 1975, 'Neoglacial Paleoenvironmental Change at Engabreen, Svartisen, Holandsfiord, North Norway', *Norges Geol. Underokelse* **321**, 37–66.

Wright, H. E.: 1984, 'Late Glacial and Late Holocene Moraines in the Cerros Cuchpanga, Central Peru', *Quatern. Res.* **21**, 275–85.

Zumbühl, H. J. and Holzhauser, H.: 1990, 'Alpengletscher in den Kleinen Eiszeit', *Geogr. Bern.* **31**, 3–36.

(Received 22 September, 1992; in revised form 12 October, 1993).

EVIDENCE FOR CLIMATIC CONDITIONS BETWEEN CA. 900-1300 A.D. IN THE SOUTHERN CANADIAN ROCKIES

B. H. LUCKMAN

Department of Geography, University of Western Ontario, London, N6A 5C2, Canada

Abstract. Available evidence for climatic conditions in the southern Canadian Rockies around the period of the Early Medieval Warm Period is presented and reviewed. Treelines appear to have been above present levels during the 14th–17th centuries and there is limited evidence of higher treelines ca. 1000 ^{14}C yr B.P. (ca. 1000 A.D.). During the 13th century at least three glaciers were advancing over mature forest in valley floor sites, 0.5–1.0 km upvalley of Little Ice Age maximum positions attained in the 18th and 19th centuries. Tree-ring width chronologies from treeline sites show suppressed growth in the early 12th century and for several periods in the 12th–14th centuries. The only tree-ring chronology presently spanning the 900–1300 A.D. interval has generally wider ringwidths between 950 and 1100 A.D. suggesting conditions were more favourable at that time. Forested sites overrun by glaciers in the 12th–14th centuries have only been deglaciated within the present century.

1. Introduction

In the Southern Canadian Rockies the broad outline of the climatic history of the last three centuries has been established by investigations of the timing and extent of glacier fluctuations and, more recently, dendrochronology (Heusser, 1956; Luckman, 1986, 1992). Most glaciers in the Canadian Rockies advanced to their maximum Holocene extent after 1700 A.D. during the Little Ice Age. These advances overran the record of earlier glacial events and, until recently, there has been little detailed information about conditions between ca. 900–1600 A.D. in this area. Available palynological records (e.g. Luckman and Kearney, 1986; Beaudoin and King, 1990; Reasoner and Hickman, 1989) generally have inadequate temporal or paleoenvironmental resolution in this time frame and knowledge of earlier glacial events is fragmentary (Osborn and Luckman, 1988). The development of long tree-ring chronologies and study of tree stumps overridden by glaciers is providing new information about this period. This paper reviews the published evidence for climatic conditions in the Canadian Rockies during the interval ca. 900–1300 A.D. that corresponds to the 'Medieval Warm Period' of Europe (Lamb, 1977; Grove, 1988). It will also present preliminary results from recent studies of tree-line fluctuations, glacier fluctuations and dendrochronology that provide new information about this time period in the Canadian Rockies.

Climatic Change **26**: 171–182, 1994.
© 1994 *Kluwer Academic Publishers.*

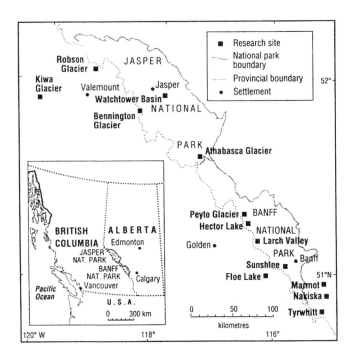

Fig. 1. Location of the Southern Canadian Rockies and research sites mentioned in the text.

2. Studies of Tree-line Fluctuations

Most reconstructions of treeline fluctuations in the Canadian Rockies have been based on palynological studies (e.g. Luckman and Kearney, 1986; Beaudoin, 1986; Reasoner and Hickman, 1989) with primary emphasis on the Holocene record and the early to mid- Holocene Hypsithermal Interval. These palynological studies and the glacial record suggest that the Hypsithermal/Neoglacial transition in this area took place ca. 4000–5000 yr ago (Osborn and Luckman, 1988; Luckman, 1990). Reconstructed treeline fluctuations for the Watchtower Basin (Jasper National Park, Luckman and Kearney, 1986), based on pollen ration data, show a short period of higher treeline just prior to ca. 1000 ^{14}C yr. B.P. (ca. 1000 A.D.; see note, Table I) but, as this high treeline phase is inferred from a single sample with a temporal resolution of ca. 200 yr/sample, it does not reveal detail about this period.

Direct evidence of former higher treelines comes from sites where standing and fallen snags (dead trees and stumps) occur at and above present treeline or where these snags are considerably larger than trees presently growing at the site. Such occurrences have been reported at four sites by Heusser (1956) and from Larch Valley near Lake Louise (Miller, 1982). Luckman (1986) dated snags at

TABLE I: Calendar and radiocarbon dates from trees at sites in the Canadian Rockies

Sample number	Ring count	Rings dated	Radiocarbon date (yr BP)	Lab #	Calendar Death Date (i)	date (ii)
Athabasca Glacier						
A80A1C* 308		170–200	1065±65	B1829	1010–1120	1691
A80D2* 198		163–198	1110±65	B1830	800–1000	1676
A79S1* 110		10–17	940±50	GSC2806	1100–1250	undated
A80E1* 118		50–75	1050±85	B1828	765–1080	undated
Robson Glacier						
Heusser[#] 135		NA	450±150	NA	1320–1640	NA
R802S*	126	1–15	860±50	B2816	1135–1345	1246
R815S*	428	48–63	1140±80	B9458	1153–1353	1319
R816S*	186	136–156	900±60	B3731	1060–1255	1253
Kiwa Glacier, Premier Ranges						
R841*	c.100	1–30	840±80	B12416	1120–1350	undated
Bennington Glacier						
R8250*	481	1–17	900±70	B12417	1020–1220**	1113**
Peyto Glacier						
GH92″	NA	NA	1710±60	B48499	240–420	undated
GDO90″	NA	outer?	1550±60	B38678	420–640	undated
PS8801″	NA	NA	1140±75	SRC2990	780–990	undated
P9140″	165	inner	1110±60	SRC3112	930–1140	undated
P9140″	165	outer	920±60	SRC3111	1040–1240	undated
P9140″	165	outer	820±50	B33010	1180–1380	undated
P9143B	662	589–631	810±70	B51229	1150–1270	1324

Notes: At the Robson, Kiwa and Peyto sites the trees dated were overridden by glaciers. The ^{14}C dates from the Athabasca site are from snags lying on the surface at treeline upslope from the Little Ice Age moraine. The date from Bennington Glacier is from a standing dead snag adjacent to the Little Ice Age moraine. Calendar death dates were determined as follows – column (i): The calendar equivalent age range for the radiocarbon dated sequence of rings is assigned to the mid-point of that ring sequence and a correction added for sample position on the snag to determine the end date; Column (ii): The crossdated calendar date of the outermost ring on the cross-section.
References * = Luckman 1986; # = Heusser 1956; ″ Luckman *et al.*, 1993; NA = not available; ** 1113 is the pith date for this tree and the calendar equivalent date is for rings 1–17.

Note on calendar equivalent dates: The calendar equivalent radiocarbon dates cited in this paper are based on Stuiver, 1982; Figure 3 (as used in Luckman, 1986, Tables III and IV). Where a single ^{14}C age is cited, an approximate calendar conversion is given. Where a ^{14}C date is cited, the range is calculated as the ^{14}C date ± 1 standard deviation. Where multiple calendar dates exist for a given ^{14}C date, the entire calendar range is given.

treeline upslope from the terminal moraine of Athabasca Glacier on the south facing slope of Wilcox Peak. Radiocarbon dates between 1110 and 940 ^{14}C yr B.P. (ca. 800–1250 A.D.) were obtained from four large snags (Table I) indicating that mature, standing trees grew at or upslope of the present treeline during this period. Hamilton (1987) used ring-width and maximum density series to produce a 300-yr floating chronology from three snags at this site (including A80A1C and A80D2, Table I) but was unsuccessful in attempts to crossdate this chronology with a 659-yr-chronology developed from living *Picea engelmannii* growing on the adjacent slope. Five other snags from the slope crossdated with this living tree chronology: earliest ring dates were between 1315 and 1446 A.D. and the trees died between 1656 and 1696 A.D. These data were interpreted as evidence of two periods of higher treeline, the floating chronology (ca. 700–1200 A.D.) corresponding to the Early Medieval Warm Period and a later phase between ca. 1350 and 1700 A.D. (Luckman, 1986) that immediately preceded the earliest dated Little Ice Age glacier advance at this site (Heusser, 1956; Luckman, 1988).

Subsequent resampling and additional crossdating has provided calendar ages for more than 90 snags at this site (Figure 2). Two of the ^{14}C-dated snags crossdate at much younger ages than indicated by the earlier radiocarbon dates (A80A1C, A80D2, Table I). Although radiocarbon dates in this time frame may have multiple calendar equivalents (Stuiver, 1982), errors of ca. 500 yr are unusual (but not unknown in the dendrochronological literature, see Baillie, 1990, 1991). Dating comparisons for trees of similar age at other sites in this area (Robson, Peyto and Bennington Glaciers, Table I) show much closer agreement between dendrochronologically-determined ages and calibrated (calendar-equivalent) radiocarbon dates.

The period of record for the snags shown in Figure 2 provides a minimum lifespan for each tree. Outer ring dates are probably close approximations to death dates, although some trees have lost outer rings due to weathering or decay. The earliest ring dates are more likely to underestimate inception dates because cross sections were often cut an unknown distance above the rootstock or because the inner rings of the tree were lost to heartrot.

These new results (Figure 2) clearly support the evidence cited above for a higher treeline phase during the 14th–17th centuries. Most snags began growing in the late 14th, 15th and early 16th centuries and died in the latter half of the 17th century (half of the snags have outer ring dates between 1658 and 1701). Evidence for an earlier high treeline stand during the 'Medieval Warm Period' is now less convincing. Only eight of the crossdated snags (Figure 2) have earliest rings prior to 1280. Attempts to crossdate two of the radiocarbon-dated snags (A79S1 and A80E1, Table I) have been unsuccessful and about 20 other snags remain undated (see Luckman *et al.*, 1992, Table 4). These crossdating results could be attributed to the generally short records in these snags (most are only 100–150 yr), unique characteristics in individual ring-width records (reaction wood, asymmetrical growth, etc.), poor replication in the early years of the site master chronology or insufficient overlap with the master chronology. Some of these snags could predate the master: the radiocarbon dates from samples A79S1 and A80E1 (Table I) suggest at least two trees were growing on the slope ca. 1000 ^{14}C years B.P. (ca. 1000 A.D.) Further sampling and crossdating trials will be necessary to confirm details of this earlier high treeline period.

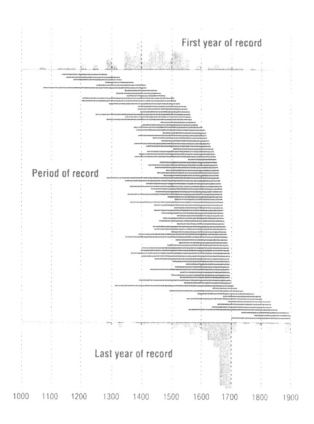

Fig. 2. Length of record preserved in calendar-dated snags recovered from the Athabasca Glacier Site. This diagram updates Luckman, 1992, Figure 4 and shows the 92 snags that had been calendar crossdated by April 1992 (Luckman *et al.*, 1992, Table 4). As explained in the text, the period of record shown is only a minimum estimate of the life of the tree. The two main Little Ice Age advances at Athabasca Glacier have been dated at 1714 and 1843–44 using ice-damaged trees (see Luckman, 1988).

Despite the partial loss of records due to rot or weathering, many of the trees in Figure 2 show a remarkably similar pattern of recruitment and mortality. These data suggest that trees colonised the slope between the late 1300s and mid-1500s and that a major dieback of trees occurred in the latter half of the 17th century. Treeline has also advanced at this site in the present century (Kearney, 1982). A smaller group of trees began growing in the late 11th and 12th centuries and some large snags remain undated. Therefore, although some evidence is present for a higher treeline stand prior to ca. 1200 A.D., most of the snags at the Athabasca site appear to date from the period ca. 1400–1700 A.D.

3. Evidence of Early Little Ice Age Glacier Fluctuations

The inception of the Little Ice Age in the Canadian Rockies was first dated by Heusser (1956) to ca. 450±150 yr. B.P. (1320–1640 A.D.) based on an early ^{14}C date from a tree overridden by Robson Glacier. Further radiocarbon dates from trees at this site (Table I) indicated that this advance probably took place somewhat earlier in the 12th and 13th centuries (Luckman, 1986). An equivalent advance occurred at Kiwa Glacier, about 50 km SW of Robson (Figure 1), where a sheared, *in situ* stump near the glacier snout was dated to 840±80 ^{14}C yr B.P. (1120–1350 A.D., Table I, see Watson, 1986; Luckman, 1986). Supporting evidence for an early Little Ice Age advance dating ca. 550–900 ^{14}C yr B.P. (ca. 1040–1400 A.D.) has been found in several other areas of the southern Canadian Cordillera (Luckman, 1986; Ryder and Thompson, 1986; Osborn and Luckman, 1988).

The buried forest site at Robson Glacier occurs 5–700 m inside the Little Ice Age terminal moraine where channels, cut through the glacial deposits, exposed *in situ* stumps and reworked detrital logs. Dendrochronological studies of this wood, using ring-width and tree-ring densitometry, led to the development of a floating chronology almost 500 yr long from cross sections of 37 logs. Calendar dates have been assigned to this material by crossdating with an 884-yr-long *Pinus albicaulis* chronology from Bennington Glacier, 70 km southeast of Robson Glacier. These dates indicate that Robson Glacier was advancing into forest between ca. 1142 and 1350 A.D. (34 of 37 kill dates are between 1214 and 1350 A.D., Luckman, unpublished data).

Recent investigations at Peyto Glacier have also recovered stumps and detrital wood from the glacier forefield that are inferred to have been killed by an advancing glacier. Radiocarbon dates between 810 and 1140 ^{14}C yr B.P. (780–1260 A.D.) have been obtained from samples of these trees (Table I; Luckman *et al.*, 1993). Calendar outer ring (kill?) dates of 1246, 1286 and 1324 A.D. have been established for three of these snags (Reynolds, 1992). Although no rooted stumps of this age have yet been found *in situ* at Peyto Glacier, this evidence probably indicates that a glacier advance was in progress more than 0.5 km upvalley from the Little Ice Age maximum limit of Peyto Glacier during the 13th and early 14th centuries.

The wood dating between 810 and 1140 yr B.P. (780–1260 A.D.) at Peyto was recovered from sites on the valley floor in the central part of the valley. Additional exposures in the lateral moraine, several hundred meters upvalley, have yielded logs dating 1550±60 (420–640 A.D.). and 1710±60 yr. B.P. (240–420 A.D., see Table I and Luckman *et al.*, 1993). These dates indicate that the glacier was advancing downvalley shortly after 1550 yr B.P. (ca 430–540 A.D.) No information is available to reconstruct glacier positions between 1500 and 900 yr B.P. (ca 500–1200 A.D.) at this site.

The radiocarbon and dendrochronological dating of trees recovered from sites within the Little Ice Age maximum limit at two sites indicates that glaciers were advancing in the Canadian Rockies during the late 12th, 13th and early 14th

TABLE II: Maximum ages of treeline species in the Canadian Rockies

Species	Maximum age	Site	Date sampled
Pinus albicaulis	> 882	Bennington Glacier	1986
Picea engelmannii	> 730	Peyto Glacier	1991
Larix lyallii	> 728	Storm Mountain	1990

The ages given are based on pith dates from positions 0.7–1.3 m above the root crown.

centuries. This corroborates the 12th–13th century glacier advance proposed by Leonard (1986a, b) based on studies of sedimentation rates in Bow and Hector Lakes. Evidence from Peyto Glacier indicates that the glacier was also advancing shortly after 1550 ^{14}C yr B.P. (ca. 500 A.D.) but no data are available to indicate glacier positions between ca. 1500 and 900 yr B.P. (ca. 500–1200 A.D.)

4. Dendrochronological Studies

The greatest potential for proxy climate records covering the period 900–1300 A.D. lies in the development of long tree-ring records from temperature-sensitive treeline sites. The earliest studies of tree-rings in the Canadian Rockies focused on trees at the lower forest/grassland border (primarily *Pseudotsuga menziesii*; e.g. Drew, 1975; Fritts *et al.*, 1979; Robertson and Jozsa, 1988). Studies of treeline sites, more directly relevant to this enquiry, began with the exploratory work of Parker and Henoch (1971) and, more recently, Schweingruber has collected from several treeline sites in this area as part of a new densitometric chronology network in the North American Cordillera (Schweingruber, 1988; Schweingruber *et al.*, 1991). Investigations at the University of Western Ontario have concentrated on developing ring-width chronologies for *Picea engelmannii* and *Larix lyallii* at treeline sites within and adjacent to Banff and Jasper National Parks (Colenutt and Luckman, 1991; Colenutt, 1992; Luckman and Colenutt, 1992; Luckman, 1992). Most chronologies are ca. 300–400 yr long but maximum living-tree ages exceed 700 yr for three treeline species (Table II). Several opportunities also exist for the development of floating chronologies from sub-fossil material.

Preliminary ring-width chronologies exceeding 800 yr have been developed for trees at sites near Bennington and Athabasca Glaciers (Figures 1–3). The trees at the Bennington site grow on a steep (20–40°) south-facing slope overlooking the Little Ice Age moraine of Bennington Glacier. An open-grown stand of *Pinus albicaulis* on this slope includes the oldest documented living trees in the Canadian Rockies (3 trees exceed 800 yr). The present chronology covers the period 1104–1989 based mainly on cores from living trees: the early part of this chronology is supplemented with ring-width records from snag material but they do not extend the living-tree chronology.

The Athabasca site is also a south-facing slope that rises above the terminal moraine of Athabasca Glacier. An old-growth stand containing *Picea engelmannii*

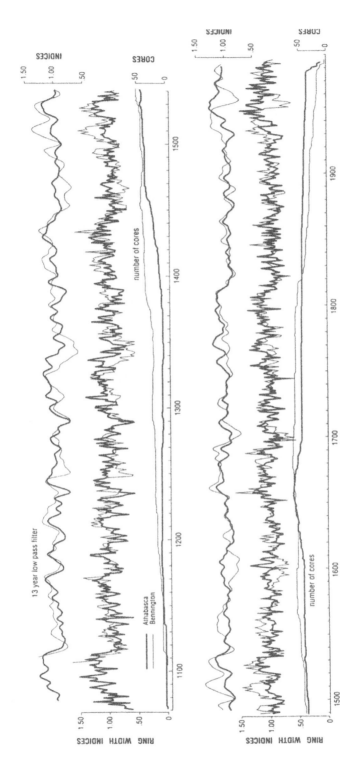

Fig. 3. Comparison of preliminary tree-ring width chronologies from the Athabasca Glacier (*Picea engelmannii*) and Bennington Glacier (*Pinus albicaulis*) sites. These chronologies are simple ARSTAN chronologies (Holmes, 1992) using only negative exponential or straight line indexing without secondary modelling to preserve the low-frequency record. The low pass filter used is given in Fritts (1976).

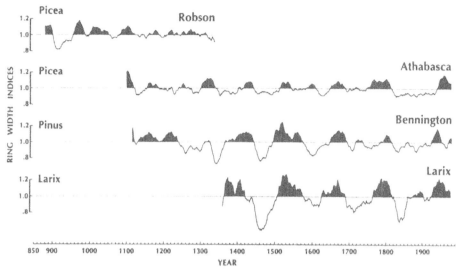

Fig. 4.   Comparison of long tree-ring width chronologies from the Canadian Rockies. The time series shown are smoothed with a 25-yr running mean and plotted at the same vertical scale: periods when tree-ring indices exceed the mean for each record are shaded. Location of the sites is given in Figure 1. The *Larix* chronology is the mean of 6 site chronologies (see text).

of up to 680 yr (Luckman *et al.*, 1984) occurs adjacent to this moraine. The chronology from this site has been extended to 1074 A.D. by crossdating snags lying on the surface near present treeline (Figure 2).

The preliminary chronologies presented in Figure 3 show strong crossdating and similarities in the ring-width records at the two sites. As chronology development is continuing, detailed attempts at paleotemperature reconstruction are premature.

Nevertheless, the synchronous variation in ring-width patterns in two species from sites almost 100 km apart provides a strong *prima facie* case for a common climatic control. Previous work on treeline spruce (e.g., Jacoby and Cook, 1981; Jacoby and D'Arrigo, 1989) and limited calibration studies (Luckman *et al.*, 1985; Colenutt and Luckman, 1991) indicate this is probably a summer temperature signal.

These records show major periods of declining ringwidth in ca. 1110–1120, possibly 1340–1360 (slightly out of phase) 1430–1460, 1690–1705 and 1800–1820 with shorter periods elsewhere in the record (e.g. 1170–1180 and 1275–1290). The declines in ringwidth at the end of 16th and in the early 19th centuries correspond with periods of known glacier advance in the Canadian Rockies (Luckman, 1986) and the early 19th century event has been recognised at latitudinal treeline sites across North America (Jacoby *et al.*, 1988; Jacoby and D'Arrigo, 1989).

Interpretations based on these long chronologies can be supplemented from shorter records at other sites. The oldest record is a snag (picea) chronology from the Robson Glacier site that covers the period 865–1350 A.D. In addition, Colenutt (1992) has developed chronologies for 6 *Larix lyallii* treeline sites (Larch Valley,

Sunshine, Floe Lake, Tyrwhitt, Marmot and Nakiska, Figure 1). The oldest trees at three of these sites exceed 600 yr, though no single chronology has great sample depth for the earliest part of the record. Figure 4 shows 25–yr running means for chronologies at the Athabasca, Bennington and Robson sites plus a composite chronology for the 6 larix sites. These chronologies are from three different treeline species at sites spaced over a 400 km-long section of the Canadian Rockies. All show strong similarities in the temporal pattern of variation after 1400 A.D. although the larix and pinus chronologies show a much greater amplitude of ring-width variation (see Luckman and Colenutt, 1992; Colenutt and Luckman, 1991). The Athabasca, Bennington and Robson sites all show a significant decrease in ringwidth in the early part of the 12th century, prior to the date that the first tree was overridden at Robson, and the longer chronologies show a growth reduction in the mid-1300s when the last Robson tree was killed.

The Robson chronology is the only one to span the entire 900–1300 period although it is only based on two trees prior to 1040. Ringwidths are generally much wider between 950 and 1100 than between 1100 and 1300. If this ringwidth record contains a proxy temperature signal it indicates that any Late Medieval Warm Period in the Canadian Rockies was probably restricted to ca. 950–1100 A.D. Initial results from Peyto Glacier indicate that tree-ring record extends back to at least 760 A.D. (Reynolds, 1992). Future chronology development at that site will assist interpretation of the Robson record.

5. Conclusion

It will be some time before a comprehensive picture of climate in the southern Canadian Rockies between ca. 900–1300 A.D. can be developed. However, the preliminary data presented here provide an intial outline of conditions during that interval.

The ringwidth records shown in Figures 3 and 4 indicate that the pattern of climate variability of the last few centuries extends back over most of the last millennium. Treelines appear to have been above present levels during the 14th–17th centuries and there is limited evidence of higher treelines ca. 1000 ^{14}C yr B.P. During the 13th century glaciers at three sites were advancing over mature forest in valley floor sites 0.5–1.0 km upvalley from LIA maximum positions. Available tree-ring width chronologies show suppressed growth in the early 12th century and for several periods in the 12th–14th centuries. The only chronology presently spanning the 900–1300 A.D. interval has generally wider ringwidths during the 950–1100 A.D. period, suggesting conditions were more favourable for growth at that time. The fact that during the 12th and 13th centuries 3 widely separated glaciers overrode mature forests at localities where present topoclimates inhibit forest development suggests that conditions during the latter part of the 10th and 11th centuries were similar to or more favourable than those during the present century.

Acknowledgements

The research reported in this paper has been supported by the Natural Sciences and Engineering Research Council of Canada. Field sampling, ring-width measurement and chronology development has been assisted by: A. B. Beaudoin, I. M. Besch, M. E. Colenutt, F. F. Dalley, S. E. Daniels, G. W. Frazer, J. P. Hamilton, R. W. Heipel, L. A. Josza, P. E. Kelly, D. C. Luckman, D. P. McCarthy, W. Quinlan, J. R. Reynolds, C. J. Rowley, D. Smith, A. Tarussov, R. Young and Yellowhead Helicopters, Valemount, B. C. I thank M. E. Colenutt and R. J. Reynolds for permission to cite unpublished data; G. Shields, U.W.O. for Cartography; Forintek Canada Corporation for Densitometric Analyses: Parks Canada, Kananaskis Country and Mount Robson Provincial Parks for permission to sample in these areas.

References

Baillie, M. G. L.: 1990, 'Checking back on an Assemblage of Published Radiocarbon Dates', *Radiocarbon* **32**, 361–366.

Baillie, M. G. L.: 1991, 'Suck-in and Smear', Two Related Chronological Problems for the 90s', *J. Theor. Archeol.* **2**, 12–16.

Beaudoin, A. B.: 1986, 'Using Picea/Pinus Ratios from the Wilcox Pass Core, Jasper National Park, Alberta, to Investigate Holocene Timberline Fluctuations', *Geogr. phys. Quatern.* **40**, 145–152.

Beaudoin, A. B. and King, R. H.: 1990, 'Late Quaternary Vegetation History of Wilcox Pass, Jasper National Park, Alberta', *Paleogeogr. Paleoclimatol. Paleoecol.* **80**, 129–144.

Colenutt, M. E.: 1992, *An Investigation of the Dendrochronological Potential of Alpine Larch*, Unpub. M. Sc. thesis, University of Western Ontario, 226 pp.

Colenutt, M. E. and Luckman, B. H.: 1991, 'Dendrochronological Studies of *Larix lyallii* at Larch Valley, Alberta', *Canad. J. Forest Res.* **21**, 1222–1233.

Drew, L. G. (ed.): 1975, *Tree-ring Chronologies of Western America, VI. Western Canada and Mexico*, Laboratory of Tree-Ring Research, Chronology Series 1, Univ. of Arizona, Tucson.

Fritts, H. C.: 1976, *Tree Rings and Climate*, Academic Press, London and New York, 568 pp.

Fritts, H. C., Lofgren, G. R., and Gordon, G. A.: 1979, 'Variations in Climate since 1602 as Reconstructed from Tree Rings', *Quatern. Res.* **12**, 18–46.

Grove, J.: 1988, *The Little Ice Age*, Cambridge Univ. Press.

Hamilton, J. P.: 1987, *Densitometric Tree-Ring Investigations at the Columbia Icefield, Jasper National Park*, M.Sc. thesis, Univ. of Western Ontario, London, Ontario, 254 pp.

Heusser, C. J.: 1956, 'Postglacial Environments in the Canadian Rocky Mountains', *Ecol. Monogr.* **26**, 253–302.

Holmes, R. W.: 1992, *Dendrochronology Program Library Installation and Program Manual* (January 1992 update), Unpub. manuscript, Tree-Ring Laboratory, Univ. of Arizona, Tucson, 35 pp.

Jacoby, G. C. and Cook, E. R.: 1981, 'Past Temperature Variations Inferred from a 400-Year Tree-Ring Chronology from Yukon Territories, Canada', *Arc. Alp. Res.* **13**, 409–418.

Jacoby, G. C. and D'Arrigo, R.: 1989, 'Reconstructed Northern Hemisphere Annual Temperature since 1671 Based on High-Latitude Tree-Ring Data from North America', *Clim. Change* **14**, 39–49.

Jacoby, G. C., Ivanciu, I. S., and Ulan, L. D.: 1988, 'A 263-Year Record of Summer Temperature for Northern Quebec Reconstructed from Tree-Ring Data and Evidence of a Major Climatic Shift in the Early 1800's', *Palaeogeogr. Palaeoclimatol. Palaeoecol.* **64**, 69–78.

Kearney, M. S.: 1982, 'Recent Seedling Establishment at Timberline in Jasper National Park', *Canad. J. Botany* **60**, 2283–2287.

Lamb, H. H.: 1977, *Climate History and the Future*, Methuen, London.

Leonard, E. M.: 1986a, 'Varve Studies at Hector Lake, Alberta, Canada, and the Relationship between Glacial Activity and Sedimentation', *Quatern. Res.* **25**, 199–214.

Leonard, E. M.: 1986b, 'Use of Lacustrine Sedimentary Sequences as Indicators of Holocene Glacial History, Banff National Park, Alberta, Canada', *Quatern. Res.* **26**, 218–231.

Luckman, B. H.: 1986, 'Reconstruction of Little Ice Age Events in the Canadian Rockies', *Geogr. phys. Quatern.* **XL**, 17–28.

Luckman, B. H.: 1988, 'Dating the Moraines and Recession of Athabasca and Dome Glaciers, Alberta, Canada', *Arc. Alp. Res.* **20**, 40–54.

Luckman, B. H.: 1990, 'Mountain Areas and Global Change – A View from the Canadian Rockies', *Mountain Research and Development* **10**, 183–195.

Luckman, B. H.: 1992, 'Glacier and Dendrochronological Records for the Little Ice Age in the Canadian Rocky Mountains', In Mikami, T. (ed.), *Proceedings of the International Conference on the Little Ice Age Climate*, Tokyo Metropolitan Univ., pp. 75–80.

Luckman, B. H. and Colenutt, M. E.: 1992, 'Developing Tree-Ring Series for the Last Millennium in the Canadian Rocky Mountains', in Bartolin, T. S., Berglund, B. E., Eckstein, D., and Schweingruber, F. H. (eds.), *Tree Rings and Environment*, Proc. of the Inter. Dendroecological Symposium, Ystad, South Sweden, 3–9 Sept. 1990. Lundqua Report 34; Dept. of Quaternary Geology, Lund Univ, pp. 207–211.

Luckman, B. H., Colenutt, M. E., and Reynolds, J. R.: 1992, 'Field Investigations in the Canadian Rockies in 1991', Report to Parks Canada, B.C. Parks and Alberta Parks, April 1992, iii + 65 pp.

Luckman, B. H., Hamilton, J. P., Jozsa, L. A., and Gray, J.: 1985, 'Proxy Climatic Data from Tree Rings at Lake Louise, Alberta: A Preliminary Report', *Geogr. phys. Quatern.* **XXXIX**, 127–140.

Luckman, B. H., Holdsworth, G., and Osborn, G. D.: 1993, 'Neoglacial Glacier Fluctuations in the Canadian Rockies', *Quatern. Res.* **39**, 144–153.

Luckman, B. H., Jozsa, L. A., and Murphy, P. J.: 1984, 'Living Seven-Hundred-Year-Old *Picea engelmannii* and *Pinus albicaulis* in the Canadian Rockies', *Arct. Alp. Res.* **16**, 419–422.

Luckman, B. H. and Kearney, M. S.: 1986, 'Reconstruction of Holocene Changes in Alpine Vegetation and Climate in the Maligne Range, Jasper National Park, Alberta', *Quatern. Res.* **26**, 244–261.

Miller, K. J.: 1982, *Vegetation Pattern and Disturbance Effects in the Subalpine-Alpine Interface, Valley of the Ten Peaks, Alberta*, Unpub. M.A. thesis, Wilfrid Laurier Univ., 256 pp.

Osborn, G. D., and Luckman, B. H.: 1988, 'Holocene Glacier Fluctuations in the Canadian Cordillera (Alberta and British Columbia)', *Quatern. Sci. Rev.* **7**, 115–128.

Parker, M. L. and Henoch, W. E. S.: 1971, 'The Use of Engelmann Spruce Latewood Density for Dendrochronological Purposes', *Canad. J. Forest Res.* **1**, 90–98.

Reasoner, M. L. and Hickman, M.: 1989, 'Late Quaternary Environmental Change in the Lake O'Hara Region, Yoho National Park, British Columbia', *Paleogeogr. Paleoclimatol. Paleoecol.* **72**, 291–316.

Reynolds, J. R.: 1992, *Dendrochronology and Glacier Fluctuations at Peyto Glacier, Alberta*, Unpub. B.A. thesis, Univ. of Western Ontario.

Robertson, E. O. and Jozsa, L. A.: 1988, 'Climate Reconstruction from Tree Rings at Banff', *Canad. J. Forest Res.* **18**, 888–900.

Ryder, J. R. and Thompson, B.: 1986, 'Neoglaciation in the Southern Coast Mountains of British Columbia: Chronology Prior to the Late-Neoglacial Maximum', *Canad. J. Earth Sci.* **23**, 273–287.

Schweingruber, F. H.: 1988, 'A New Dendroclimatic Network for Western North America', *Dendrochronol.* **6**, 171–180.

Schweingruber, F. H., Briffa, K. R., and Jones, P. D.: 1991, 'Yearly Maps of Summer Temperatures in Western Europe from A.D. 1750 to 1975 and Western North America from 1600 to 1982', *Vegetatio* **92**, 5–71.

Stuiver, M.: 1982, 'A High Precision Calibration of the A.D. Radiocarbon Time Scale', *Radiocarbon* **24**, 1–26.

Watson, H. W.: 1986, *Little Ice Age Glacial Fluctuations in the Premier Ranger, B.C.*, Unpub. M.Sc. thesis, Univ. of Western Ontario.

(Received 22 September, 1992; in revised form 12 October, 1993)

TREE-RING AND GLACIAL EVIDENCE FOR THE MEDIEVAL WARM EPOCH AND THE LITTLE ICE AGE IN SOUTHERN SOUTH AMERICA

RICARDO VILLALBA*

Department of Geography, University of Colorado, Boulder, CO 80309-260, U.S.A.

Abstract. A tree-ring reconstruction of summer temperatures from northern Patagonia shows distinct episodes of higher and lower temperature during the last 1000 yr. The first cold interval was from A.D. 900 to 1070, which was followed by a warm period A.D. 1080 to 1250 (approximately coincident with the *Medieval Warm Epoch*). Afterwards a long, cold-moist interval followed from A.D. 1270 to 1660, peaking around 1340 and 1640 (contemporaneously with early *Little Ice Age* events in the Northern Hemisphere). In central Chile, winter rainfall variations were reconstructed using tree rings back to the year A.D. 1220. From A.D. 1220 to 1280, and from A.D. 1450 to 1550, rainfall was above the long-term mean. Droughts apparently occurred between A.D. 1280 and 1450, from 1570 to 1650, and from 1770 to 1820. In northern Patagonia, radiocarbon dates and tree-ring dates record two major glacial advances in the A.D. 1270–1380 and 1520–1670 intervals. In southern Patagonia, the initiation of the *Little Ice Age* appears to have been around A.D. 1300, and the culmination of glacial advances between the late 17th to the early 19th centuries.

Most of the reconstructed winter-dry periods in central Chile are synchronous with cold summers in northern Patagonia, resembling the present regional patterns associated with the El Niño-Southern Oscillation (ENSO). The years A.D. 1468–69 represent, in both temperature and precipitation reconstructions from tree-rings, the largest departures during the last 1000 yr. A very strong ENSO event was probably responsible for these extreme deviations. Tree-ring analysis also indicates that the association between a weaker southeastern Pacific subtropical anticyclone and the occurence of El Niño events has been stable over the last four centuries, although some anomalous cases are recognized.

1. Introduction

Paleoclimatic studies in the Southern Hemisphere have historically lagged those of the Northern Hemisphere. Recent Holocene records for the Southern Hemisphere, (particularly from South America), are comparatively rare. However, for a global interpretation of climate variations at any time scale, there is a need to examine and incorporate the records from the Southern Hemisphere.

In the Northern Hemisphere, the two most significant climatic events recognized during the last millennium are the *Medieval Optimum* and the *Little Ice Age*. The evidence for the *Medieval Optimum* period is mainly from Europe and the North Atlantic (Lamb, 1977; Williams and Wigley, 1983). A clear definition for the timing

* Present address: Laboratorio de Dendrocronologia, CRICYT - CONICET, Casilla de Correo 330, 5500 Mendoza, Argentina.

Climatic Change **26**: 183–197, 1994.

Fig. 1.   Location map of tree-ring and glaciological records in southern South America.

and character of this period is not available on a global basis. Historical evidence of the *Little Ice Age* events is much more complete in Europe than elsewhere, but its expression in other continents is supported by much field evidence (Grove, 1988).

Southern South America (Figure 1), the only continental land mass in the Southern Hemisphere extending as far south as 55°S, presents a unique opportunity to reconstruct terrestrial records of paleoclimate in a region that is under the influence of polar and mid-latitude atmospheric circulation features (Taljard, 1972). This paper provides an overview of the most significant climatic fluctuations in southern South America during the last 1000 yr derived from tree-ring records and other paleoclimate indicators. Because no global consensus exits for the dates of the *Medieval Optimum* and the *Little Ice Age* events, Lamb (1977) dates, largely based on European evidences, are used here as temporal references only.

Climatic studies using modern instrumental records (i.e. over the past c. 100 yr) indicate that the South American climate is strongly affected by warm and cold events associated with the Southern Oscillation (SO) (Aceituno, 1988; Kiladis and Diaz, 1989). Consistent changes in the regional patterns of temperature and precipitation in South America associated with SO, provide the climatological

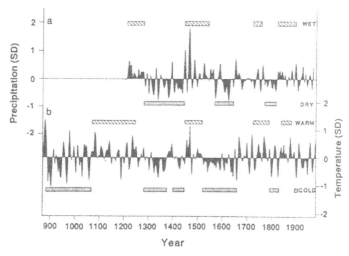

Fig. 2. Reconstructed climatic parameters derived from tree-ring records from southern South America. (a) Reconstruction of Santiago de Chile winter precipitation (Boninsegna, 1988); (b) Reconstruction of summer temperature in northern Patagonia (Villalba, 1990a). Tree-ring-based records are plotted in smoothed form to emphasize low-frequency variance (13 weight symmetrical low-pass filter; Fritts, 1976). Dashed/shaded bars indicate wet/dry and warm/cold intervals inferred from precipitation and temperature reconstructions, respectively.

basis for comparing paleoclimatological records from different regions in South America. An attempt to correlate different paleo-records in southern South America based on the modern patterns of the SO is presented. However, it is important to understand that the SO explains only part of the climate variability of southern South America. Although the current analysis focuses on the SO, other alternative sources of climatic variability could be related to the paleoclimate changes recorded.

2. Tree-Ring Record

Few tree-ring records in southern South America cover all of the last 1000 yr. Winter rainfall variations for Santiago, Chile (33°S) were reconstructed back to the year A.D. 1220 using a tree-ring chronology of *Austrocedrus chilensis*, from El Asiento, Aconcagua province, Chile (Boninsegna, 1988). The reconstruction accounts for 41% of the total variance over the calibration period. From A.D. 1220 to 1280, coinciding with the latter part of the *Medieval Optimum*, and from A.D. 1450 to 1550, rainfall was above the long-term mean (Figure 2). One long period of drought apparently occurred between A.D. 1280 and 1450, and two other drought periods appear from 1570 to 1650, and from 1770 to 1820. These winter rainfall variations are representative of precipitation variations in central Chile from 30 to 36°S. Dry winters are better reconstructed by the model than wet winters (Boninsegna, 1988).

Millennium-old *Fitzroya cupressoides* trees, located in the Río Alerce valley, Río Negro province, Argentina (41°S), were used to develop a 1120-yr reconstruction of summer temperatures (Villalba, 1990a; Figure 2). The variance in summer temperature explained by ring-width variations was slightly over 41%. The model estimates warm summers better than cold (Villalba, 1990a). The reconstruction of the summer temperature departure for northern Patagonia shows distinct episodes of higher and lower temperature during the last 1000 yr. The first cold interval was from A.D. 900 to 1070, which was followed by a warm period A.D. 1080 to 1250 (approximately coincident with the *Medieval Warm Epoch* of Europe; Lamb, 1977). Afterwards, a long, cold interval followed from A.D. 1270 to 1660, peaking around 1340 and 1640 (contemporaneously with *Little Ice Age* events in the Northern Hemisphere; Lamb, 1977). This cold interval was interrupted by a relatively short warm period between A.D. 1380 and 1520. Warmer conditions then resumed between A.D. 1720 and 1790. The mean temperature departure for the coldest interval (A.D. 1520 to 1660) is estimated to be 0.26°C lower than that for the warmest interval (A.D. 1080–1250). The temperature difference between these two intervals is significant at 99.99% level of significance (two-tailed test; Fritts, 1976). Considering that the model underestimates the real value of cold years, the temperature differences between these two intervals would be probably greater. Correlation between the Río Alerce reconstruction and a set of regional weather stations indicates that the tree-ring variations at this site are correlated with a homogeneous summer temperature pattern covering Patagonia east of the Andes from 38° to 50°S (Villalba, 1990a).

Comparison of the reconstruction of winter precipitation from central Chile with the summer temperature reconstruction from northern Patagonia, indicates that most of the winter-dry periods in central Chile are synchronous with cold summer temperatures in northern Patagonia (Figure 2). For example, during the periods A.D. 1280 to 1380 and 1570 to 1660, precipitation in central Chile was below the series mean, and northern Patagonia recorded the coolest summers of the last 1000 yr; conversely, from A.D. 1450 to 1520, from 1720 to 1760, and from 1850 to 1891, both rainfall in central Chile and Summer temperatures in North Patagonia were above the respective mean values. These climatic patterns resemble the present regional weather patterns associated with El Niño-Southern Oscillation events (Aceituno, 1988; Kiladis and Diaz, 1989).

3. Glacial Record

In central Chile, radiocarbon dates indicate advances of Los Cipreses Glacier (34°S) at the beginning of the fourteenth century and more recently around the 1860s (Röthlisberger, 1986). At a latitude of 41°S, radiocarbon dates of tree trunks overriden by the Río Manso Glacier indicate major glacial advances in A.D. 1040, 1330, 1365, 1640 and during the period A.D. 1800–1850 (Röthlisberger, 1986; Figure 3, and Table I). For the nearby Frías Glacier (41°10'S), two major glacial

TABLE I: Radiocarbon dates related to glacier advances in southern South America

No.	Date (years B.P.)	Location	Reference
1	Modern	Glaciar Los Cipreses, Chile (34° S)	Röthlisberger (1986)
2	625±155	Glaciar Los Cipreses, Chile (34° S)	Röthlisberger (1986)
3	Modern	Glaciar C. Overo, Argentina (41° S)	Röthlisberger (1986)
4	Modern	Ventisquero Negro, Argentina (41° S)	Röthlisberger (1986)
5	300±85	Ventisquero Negro, Argentina (41° S)	Röthlisberger (1986)
6	620±50	Ventisquero Negro, Argentina (41° S)	Röthlisberger (1986)
7	585±50	Ventisquero Negro, Argentina (41° S)	Röthlisberger (1986)
8	940±110	Ventisquero Negro, Argentina (41° S)	Röthlisberger (1986)
9	670±85	Glaciar Ofhidro, Chile (48° S)	Mercer (1970)
10	1180±175	Ventisquero O'Higgins, Chile (48° S)	Röthlisberger (1986)
11	165±50	Ventisquero Bravo, Chile (48° S)	Röthlisberger (1986)
12	270±100	Ventisquero Bravo, Chile (48° S)	Röthlisberger (1986)
13	665±80	Ventisquero Bravo, Chile (48° S)	Röthlisberger (1986)
14	240±50	Ventisquero Huemul, Chile (48° S)	Röthlisberger (1986)
15	310±80	Ventisquero Huemul, Chile (48° S)	Röthlisberger (1986)
16	425±45	Ventisquero Huemul, Chile (48° S)	Röthlisberger (1986)
17	465±65	Ventisquero Huemul, Chile (48° S)	Röthlisberger (1986)
18	470±55	Ventisquero Huemul, Chile (48° S)	Röthlisberger (1986)
19	895±85	Ventisquero Huemul, Chile (48° S)	Röthlisberger (1986)
20	205±50	Ventisquero Perro, Chile (51° S)	Röthlisberger (1986)
21	260±50	Ventisquero Perro, Chile (51° S)	Röthlisberger (1986)
22	295±75	Ventisquero Perro, Chile (51° S)	Röthlisberger (1986)
23	345±100	Ventisquero Perro, Chile (51° S)	Röthlisberger (1986)
24	795±80	Ventisquero Perro, Chile (51° S)	Röthlisberger (1986)
25	945±50	Ventisquero Torres, Chile (51° S)	Röthlisberger (1986)
26	235±65	Ventisquero Frances, Chile (51° S)	Röthlisberger (1986)
27	675±45	Ventisquero Frances, Chile (51° S)	Röthlisberger (1986)

advances are indicated for the A.D. 1270–1380 and 1520–1670 intervals (Villalba *et al.*,1990). However, only the latter advance has been adequately dated using tree-ring and historical records. In the South Patagonian icefield (48°S) the initial stages of Mercer's *Third Neoglacial* events are dated about A.D. 1300. In the same area, assuming a delay of c. 70 yr between deglaciation and tree establishment, the culmination of glacial advances has been estimated between the late 17th and the early 19th century (Mercer, 1968, 1970).

Most radiocarbon dates are centered around three periods: from A.D. 1280 to 1460, from 1560 to 1690, and around 1860. Radiocarbon dates 8, 19, and 25 in Table I, appear to be related to a glacier advance at the end of the 11th century. With the exception of radiocarbon date 24, no other dates suggest a glacier advance between A.D. 1100 and 1280 (Figure 3). Other intervals in which glacial recession

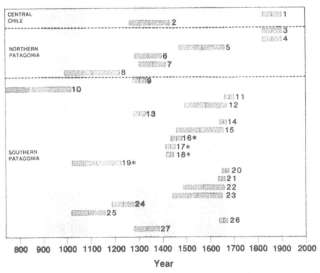

Fig. 3.   A compilation of radiocarbon dates associated with glacier advances along the Andes
ranges of southern South America during the last millennium. Radiocarbon dates are listed in Table I
according to the number assigned in this Figure. Shaded intervals correspond to cool summer periods
in northern Patagonia reconstructed from tree-rings (Villalba, 1990a). Radiocarbon dates have been
calibrated following Stuiver and Becker (1986). Boxes represent radiocarbon dates (central line) with
their respective standard deviations. An asterisk (*) indicates fossil tree-trunks not *in situ*.

probably occurred are: from A.D. 1440 to 1560, from 1700 to 1800, and from
1850 to 1900. Since the mid-19th century, a clear trend of glacier recession has
occurred in most of the glaciers along the Andes (Clapperton, 1983; Rabassa and
Clapperton, 1990).

When the glacial record is compared with the historical record of El Niño
events (Quinn and Neal, 1992; stronger events: M+ to VS intensity levels), glacial
advances and retreats appear to be associated with periods of low and high recurrence of El Niño events, respectively. Certainly, the most unusual changes in the El
Niño event recurrence occur from the A.D. 1620–1680 period, with 7 events (none
very strong), to the A.D. 1680–1740 period, with 13 events, 2 of which were very
strong El Niño events. A similar change in the event frequency is also observed
from the A.D. 1830–1860 period, with 3 events (none very strong), to the 1864–
1891 period, with 8 events, two of which were very strong events. Quinn *et al.*
(1987), and Quinn and Neal (1992) indicate that the latter change in El Niño event
frequency may have been due to the general warming that followed the late phase
of the *Little Ice Age*. The relationship between glacial advances and low frequency
changes of El Niño events appears to be valid for most of the temperate glaciers
in Patagonia (Villalba *et al.*, 1990), however, the response may be different in subtropical South America. In effect, in the dry Andes of Argentina and central Chile
(30 to 35°S), glacial advances mainly respond to an increase in annual precipitation
rather than a decrease in temperature (Leiva *et al.*, 1989). According to the modern
patterns related to the Southern Oscillation, precipitation in central Chile is higher

during the warm events, and therefore in this region glacial advances would have occurred during periods of high recurrence of El Niño events.

4. Southern South America and the Southern Oscillation

Heavy winter rainfall in central Chile is associated with positive anomalies during the developing stage of warm events of the Southern Oscillation (SO). Conversely, most cold events correspond to dry conditions (Rutllant and Fuenzalida, 1991). On the other hand, in northern Patagonia positive departures of summer temperature follow the warm events of the SO (Kiladis and Diaz, 1989). Warm and cold events of the SO (after Kiladis and Diaz, 1989) and El Niño events (from Quinn and Neal, 1992) were compared simultaneously with both the winter rainfall reconstruction in central Chile and the summer temperature reconstruction in northern Patagonia. Only the years of simultaneous positive and negative departures were tested. For the A.D. 1877–1972 interval, based on a total of 22 yr in which both series show simultaneously positive departures from the mean values, 17 years are related to warm events of the SO (Table II). A similar analysis between 1886 and 1972 indicates that from a total of 25 yr in which both series show negative departures from the mean, 18 cases can be associated with cold events of the SO. These relationships between tree-ring departures and the SO events are significant at the 95% level of significance (sign test; Fritts, 1976).

Based on the previous analysis, the occurence of a warm event of the Southern Oscillation in years in which both reconstructions show positive departures (above average winter rainfall in central Chile, and above average summer temperature in northern Patagonia) is highly probable. Conversely, negative departures in both series could indicate years of cold events. However, due to the autocorrelation in both temperature and precipitation reconstructions, it is impossible to establish if two or more consecutive departures of the same sign respond to a single, or more than one SO event. Therefore, the number of positive (negative) departures by decade is only a relative measure of the dominance of warm (cold) events than the actual frequency of warm (cold) events. Examination of both reconstructions shows the occurrence of a relatively high number of years with high rainfall in central Chile and warm summers in Patagonia in the following decades: A.D. 1240, 1460, 1520, 1740, and 1870 (Figure 4). On the other hand, the decades which record high number of years with simultaneous dry winters in central Chile and cold summers in northern Patagonia are A.D. 1360, 1470, and 1800 (Figure 4).

The years A.D. 1468–69 represent, in both reconstructions, the highest departures from their respective averages during the last 1000 yr. A warm event of the SO, larger in magnitude to those historically recorded, is probably responsible for these extreme deviations (Figure 5). Another important warm event may also be associated with higher departures in the years A.D. 1395–96. Coincidentally, corrected radiocarbon dates from detrital wood contained in flood sediments in the

TABLE II: Tree-ring reconstruction deviations and El Niño-Southern Oscillation (ENSO). Simultaneous positive and negative departures of winter rainfall in central Chile and summer temperature in northern Patagonia in relation to warm (dots) and cold (diamond) events of the El Niño-Southern Oscillation, respectively. Singles, and groups of 2 or 3 yr, associated with an ENSO event are indicated by boxes

Year	Sum. temp.	Win. prec.	ENSO	Year	Sum. temp.	Win. prec.	ENSO	Year	Sum. temp.	Win. prec.	ENSO
1877	+	+	•	1910	+	−		1943	+	−	•
1878	+	+	•	1911	+	−		1944	+	+	
1879	−	−		1912	+	+	•	1945	−	−	
1880	+	+	•	1913	+	−		1946	−	−	
1881	+	+		1914	+	+	•	1947	−	−	
1882	+	−		1915	−	−		1948	+	+	
1883	−	+		1916	−	−	◊	1949	−	−	◊
1884	−	+		1917	−	−		1950	−	+	
1885	−	+		1918	+	+	•	1951	+	−	
1886	−	−	◊	1919	+	+	•	1952	+	−	
1887	+	+	•	1920	−	+		1953	−	+	
1888	+	+	•	1921	−	+		1954	+	−	
1889	−	−	◊	1922	+	+		1955	−	+	
1890	−	+		1923	−	+		1956	+	+	
1891	+	+	•	1924	−	−	◊	1957	−	−	
1892	+	−		1925	−	+	•	1958	+	−	•
1893	+	−		1926	+	+	•	1959	+	+	
1894	+	−		1927	+	+		1960	+	−	
1895	−	−		1928	−	+		1961	+	+	
1896	+	−		1929	−	+		1962	0	+	
1897	+	−		1930	+	+	•	1963	−	+	
1898	−	+		1931	−	+	◊	1964	−	−	◊
1899	−	+		1932	−	−		1965	−	+	
1900	−	+		1933	−	−		1966	+	−	
1901	+	−		1934	−	−		1967	+	−	
1902	−	+		1935	+	−		1968	−	−	
1903	−	−	◊	1936	+	−		1969	−	−	
1904	−	+		1937	+	+		1970	−	+	
1905	−	+		1938	−	−	◊	1971	−	+	
1906	−	−	◊	1939	−	−		1972	+	+	•
1907	−	−		1940	−	+					
1908	−	−	◊	1941	−	+					
1909	−	−		1942	−	+					

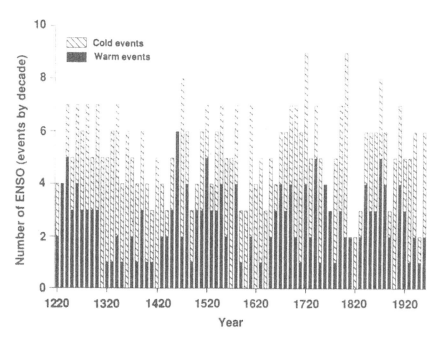

Fig. 4. Tree-ring reconstruction departures from the mean values and the El Niño-Southern Oscillation (ENSO) events. Number of years by decade with heavy winter rainfall (year 0) in central Chile concurrent with subsequent warm summers (year +1) in northern Patagonia (black bars), and low winter rainfall in central Chile concurrent with subsequent cool summer in northern Patagonia (dashed bars). Black and dashed bars are inferred to represent relative frequency of warm and cold ENSO events, respectively. See text for more details.

northern coast of Perú, also indicate the occurrence of strong El Niño events in A.D. 1460 ± 20, and A.D. 1380 ± 40 (Wells, 1987; event 11, Wells, 1990).

Heavy winter rainfall in central Chile occurs normally in El Niño years (Rutlland and Fuenzalida, 1991). They are both consequences of a weakeness in the southeastern Pacific subtropical anticyclone and remote forcing of extratropical systems by the anomalous tropical convection (Karoly, 1989). However, while high rainfall in the arid coastal lowlands of Perú results from the invasion of warm air masses from the north and west, precipitation in central Chile is related to a northward shift of the westerly storm tracks along the coast of Chile (Quinn and Neal, 1992; Aceituno, 1988).

A set of 17 tree-ring chronologies in Argentina and Chile were used to estimate the past position of the southeastern Pacific subtropical anticyclone (Villalba, 1990b). Eigenvector analysis of these chronologies revealed three dominant patterns of year-to-year tree-ring width variability. The third eigenvector, grouping the *Austrocedrus chilensis* chronologies between 39° and 42°S east of the Andes, is mainly related to the position of the anticyclone in summer. Eight of the ten most extreme northern positions, representing the weakest-index years of the southeast-

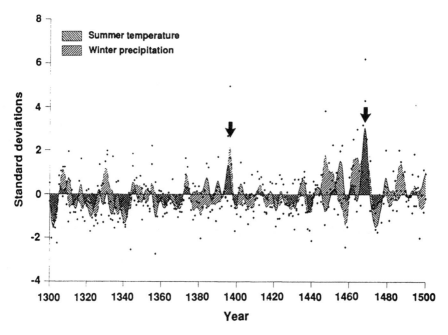

Fig. 5. Winter rainfall deviations in central Chile and summer temperature deviations in northern Patagonia for the interval A.D. 1300–1500. Note the extreme departures for the years 1395–96 and 1468–1469 (arrows). It is suggested that these departues are related to the occurrence of very strong El Niño events during those years. Actual precipitation and temperature departures are indicated by dots and diamonds, respectively.

ern Pacific anticyclone in summer, are associated with the El Niño events proposed by Quinn and Neal (1992). Two years (A.D. 1828 and 1926) are classified by these authors as very strong (VS) events, five years (A.D. 1589, 1600, 1652, 1940 and 1941) as strong (S or S+) events, and one year (1868) is indicated as a moderate (M+) event.

Winter rainfall variability in central Chile has been related to changes in the position and intensity of the adjacent subtropical anticyclone. An anticyclone displaced to the south is associated with relative dry conditions, while wet winters tend to occur when the anticyclone is shifted northward from its average latitudinal position (Rubin, 1955; Pittock, 1980; Minetti and Sierra, 1989). Consequently, the winter precipitation reconstruction of Santiago de Chile (Boninsegna, 1988) was adopted as representative of the winter anticyclone positions.

In general an association between a weaker southeastern Pacific anticyclone and the occurrence of the El Niño events is apparent (Table III). Comparison of the tree-ring estimates with the historical El Niño events (Quinn and Neal, 1992), shows that the Pacific anticyclone was displaced to the north (i.e. weaker subtropical pressure) in 67 and 75% of El Niño years for summer and winter seasons, respectively. Both relationships are significant at 99% level of significance (two-tailed test, Fritts,

TABLE III: The southeastern subtropical Pacific anticyclone and El Niño events. Comparison of the tree-ring estimated summer (a, Northern Patagonia) and winter (b, Central Chile) positions of the southeastern subtropical Pacific anticyclone with El Niño events (after Quinn and Neal, 1992). During El Niño years, the Pacific anticyclone was displaced to the north (i.e. weaker pressure) in 67 and 75% of the cases for summer and winter seasons, respectively. Years (shown in parenthesis) with anomalous anticyclone positions (in terms of the 'expected' behavior) are indicated by century and event strength. An asterisk indicates anomalous years (in terms of the 'expected' behavior), followed by normal years. If these years were considered as normal years (due to the autocorrelation in the tree ring estimates) the Pacific anticyclone would have been shifted north in 82 and 88% of El Niño years for summer and winter seasons, respectively. A.: Number of agreements by century. D.: Number of disagreements by centuries

| Century | (a) Northern Patagonia El Niño events | | | | Total number of | |
	Very Strong	Strong plus	Strong	Moderate plus	Agree-ments	Disagree-ments
16th A.	–	1	5	1	7	
D.	1(1578)	–	–	2(1565*,1585*)		3
17th A.	–	2	7	3	12	
D.	–	1(1624)	3(1614,1660*, 1681)	1(1604)		5
18th A.	2	–	4	4	10	
D.	1(1791*)	2(1701,1747)	2(1736,1783*)	2(1718*,1778–9)		7
19th A.	2	3	2	6	13	
D.	1(1891*)	1(1871*)	1(1864)	4(1817,1819, 1832*,1897*)		7
20th A.	1	–	5	4	10	
D.	–	–	1(1917*)	3(1939*,1943, 1953*)		4
Total A.	5	6	23	18	52(67%)	
D.	3	4	7	12		26(33%)

| Century | (b) Central Chile El Niño events | | | | Total number of | |
	Very Strong	Strong plus	Strong	Moderate plus	Agree-ments	Disagree-ments
16th A.	–	1	7	4	12	
D.	1(1578*)	–	–	1(1596)		2
17th A.	–	2	7	2	11	
D.	–	1(1624)	3(1671*,1681, 1692–3*)	2(1604*,1684)		6
18th A.	2	2	4	5	13	
D.	1(1791*)	–	2(1715–6,1736*)	1(1778–9)		4
19th A.	2	4	3	6	15	
D.	1(1828*)	–	–	4(1812,1817, 1819,1866*)		5
20th A.	1	–	4	5	10	
D.	–	–	2(1932,1957–8*)	2(1910,1943*)		4
Total A.	5	9	25	22	61(75%)	
D.	3	1	7	10		21(25%)

1976). This indicates the existence, during the last four centuries, of an atmospheric pattern similar to that observed today (Aceituno, 1988; Rutllant and Fuenzalida, 1991). However, some anomalies in terms of the 'expected' behavior need to be discussed. During the very strong A.D. 1578 El Niño event, both summer and winter estimates are far below the mean value, suggesting a regional southeastern Pacific anticyclone stronger than the normal. In particular, the year A.D. 1791, identified by Quinn and Neal (1992) as a very strong event (but just strong by Hocquenghem and Ortlieb, 1992), was not only climatically anomalous in central Chile but also in the Pampas of Argentina, where interannual rainfall variability is related to the El Niño-Southern Oscillation (Ropelewski and Halpert, 1987). According to Taulis (1934), precipitation in central Chile was below average (extremely dry) in 1791. In contrast to the modern patterns of the SO, Politis (1984) indicated that the year 1791 represented the most extreme dry episode during the 17th and 18th centuries in the Argentinean Pampas. Aceituno and Montecinos (1992) noted that the relationship between the SO and the interannual rainfall variability, and hence the strength of the southeastern Pacific anticyclone, has varied significantly in South America during the present century. These proxy records suggest that such variations in the relationship between ENSO and South American precipitation are part of the longer-term functioning of the ENSO system.

5. Discussion and Conclusions

In Patagonia, both tree-ring records and records of glacial variations indicate an interval of above average temperatures for A.D. 1080 to 1250. For this interval, the summer temperature reconstruction for northern Patagonia shows an increase in the frequency of warm years (without any obvious change in the intensity of warm years), and a decrease in the intensity of cold years. The whole interval is only interrupted by one cold period around A.D. 1190 (Figure 2). Interestingly, a similar cold event around A.D. 1190 has recently been reported for Tasmania, at the same latitude as northern Patagonia, by Cook et al. (1991). Below average temperatures occurred in the A.D. 1300 to 1380 and 1520 to 1660 intervals. During these cold summer periods in northern Patagonia, central Chile recorded the most intense droughts of the last 1000 yr.

In general, a tendency towards the simultaneous occurrence of positive (negative) departures of winter rainfall in central Chile and summer temperatures in northern Patagonia and warm (cold) events of the Southern Oscillation is apparent. Some departures from this general pattern could be explained. Positive departures both in central Chile precipitation and northern Patagonia summer temperatures occurred in 1948 (Table II). Even though this year is not listed in Quinn and Neal (1992) or in Kiladis and Diaz (1989) as a warm event, monthly sea surface temperatures at Puerto Chicama, Perú, exceed the mean annual cycle during the first half of 1948 by 1°C (Deser and Wallace, 1987). According to Rutlland and Fuenzalida

(1991), some dry winters in central Chile, not listed as cold events, showed anomalies distinctive of cold events. For example, during the dry 1968 winter in central Chile (followed also by a cold summer in northern Patagonia), the first half of the year presented negative departures in sea level pressure at Darwin and in the sea surface temperature at Puerto Chicama. Consequently, if this consideration is taken into account, the proposed relationships between simultaneous positive (negative) departures of rainfall and temperature in central Chile and northern Patagonia, respectively, and warm (cold) events of the SO appears highly consistent.

Kiladis and Diaz (1989), noted that during the year preceding the development of a warm event in the Southern Oscillation, climatic anomalies tend to be opposite to those during the following year. This biennial tendency of the Southern Oscillation is poorly reconstructed using *Fitzroya cupressoides* chronologies due to the high autocorrelation present in most of the tree-ring chronologies constructed for this species (Table II). Therefore, the total number of events (warm and cold) is a better indicator of extreme events in the Southern Oscillation than the frequency of warm or cold events considered separately. The periods A.D. 1240–1349, 1450–1489, 1510–1589, 1670–1759, 1780–1809, and 1840–1889, were intervals of high recurrence of the Southern Oscillation events (Figure 4). On the other hand, low event recurrence occurred from A.D. 1350 to 1449, from 1590 to 1669, and from 1810 to 1839. Most of these periods of high recurrence are also recognized by Quinn and Neal (1992).

During the 13th century (concurrent with the *Medieval Warm Period*), precipitation in central Chile and summer temperature in northern Patagonia were above average (Figure 2). If the same climatic anomalies observed today are responsible for the paleoclimatic changes reconstructed, warm-type events of the Southern Oscillation should have predominated over cold-type events during the *Medieval Warm Period*. For the Magdalena-Cauca-San Jorge river system, Van der Hammen (1991) recorded a period of low effective rainfall between 750 and 650 B.P. Low precipitation in northwestern South America is related to warm-type events of the Southern Oscillation (Aceituno, 1988). This suggests, in accordance with paleoclimate indicators from southern South America, an intensification of the warm-type events during the thirteenth century. Conversely, the prevalence of negative departures of winter rainfall and summer temperature from A.D. 1280 to 1380, and from A.D. 1520 to 1650, simultaneous with the *Little Ice Age*, could be related to a predominance of cold- over warm-type events of the SO.

As has been mentioned, the most significant patterns of climatic variations associated with the SO are relatively well known (Aceituno, 1988; Kiladis and Diaz, 1989; Rutlland and Fuenzalida, 1991). However, they only explain part of the total variance of the climate in southern South America. Other tropical and extra-tropical atmospheric-oceanic components of the climate system play an important role in climate changes in South America (COHMAP members, 1988). They could in part be responsible for the paleoclimate changes recorded in South America. Consequently, more reliable sets of paleodata for southern South America are needed in order to aid our understanding of past and present atmospheric changes, and to help us obtain a clearer regional climate chronology for the last 1000 yr.

Acknowledgements

Partial support for this study was provided by NASA Global Change Program. For review and critical comments I thank Keith Briffa, University of East Anglia, U.K., Henry Diaz, NOAA, Boulder, Colorado, Vera Markgraf, INSTAAR, and Thomas T. Veblen, University of Colorado. This is a contribution to the Project 341 IGCP/IUGS/UNESCO: Southern Hemisphere Paleo- and Neoclimates.

References

Aceituno, P.: 1988, 'On the Functioning of the Southern Oscillation in the South American Sector. Part I: Surface Climate', *Mon. Wea. Rev.* **116**, 505–524.
Aceituno, P. and Montecinos, A.: 1992, 'Análisis de la Estabilidad de la Relación entre la Oscilación Sur y la Precipitación en América del Sur', Ortlieb, L. and Macharé, J. (eds.), *Paleo ENSO Records* Intern. Symp. Extended Abstracts, ORSTOM-CONCYTEC, Lima, pp. 7–13.
Boninsegna, J. A.: 1988, 'Santiago de Chile Winter Rainfall since 1220 as Being Reconstructed by Tree Rings', *Quatern. South Amer. Antarct. Penins.* **6**, 67–87.
Clapperton, C. M.: 1983, 'The Glaciation of the Andes', *Quatern. Sci. Rev.* **2**, 83–155.
COHMAP members: 1988, 'Climatic Changes of the last 18,000 Years: Observations and Model Simulations', *Science* **241**, 1043–1052.
Cook, E., Bird, T., Peterson, M., Barbetti, M., Buckley, B., D'Arrigo, R., Francey, R., and Tans, P.: 1991, 'Climatic Change in Tasmania Inferred from a 1089-Year Tree-Ring Chronology of Huon Pine', *Science* **253**, 1266–1268.
Deser, C. and Wallace, J. M.: 1987, 'El Niño Events and Their Relation to the Southern Oscillation: 1925–1986', *J. Geophys. Res.* **92**, 14189–14196.
Grove, J. M.: 1988, *The Little Ice Age*, Routledge, London.
Fritts, H. C.: 1976, *Tree Rings and Climate*, Academic Press, London.
Hocquenghem, A. M. and Ortlieb, L.: 1992, 'Historical Record of El Niño Events in Perú (XVI-XVIIIth Centuries): The Quinn *et al.* (1987) Chronology Revisited', in Ortlieb, L. and Macharé, J. (eds.), *Paleo ENSO Records*, Extended Abstracts, ORSTOM-CONCYTEC, Lima, pp. 143–149.
Karoly, D. J.: 1989, 'Southern Hemisphere Circulation Features Associated with El Niño-Southern Oscillation Events', *J. Clim.* **2**, 1239–1252.
Kiladis, G. N. and Diaz, H. F.: 1989, 'Global Climatic Anomalies with Extremes in the Southern Oscillation', *J. Clim.* **2**, 1069–1090.
Lamb, H. H.: 1977, *Climate: Present, Past and Future*, Vol. 2, Methuen, London.
Leiva, J. C., Lenzano, L. E., Cabrera, G. A., and Suarez, J. A.: 1989, 'Variations of Río Plomo Glaciers, Andes Centrales Argentinos', in Oerlemans, J. (eds.), *Glacier Fluctuations and Climatic Change*, Kluwer Academic Publishers, pp. 143–151.
Mercer, J. H.: 1968, 'Variations of Some Patagonian Glaciers since the Late-Glacial: I.', *Amer. J. Sci.* **266**, 91–109.
Mercer, J. H. : 1970, 'Variations of Some Patagonian Glaciers since the Late-Glacial: II.', *Amer. J. Sci.* **269**, 1–25.
Minetti, J. L. and Sierra, E. M.; 1989, 'The Influence of General Circulation Patterns on Humid and Dry Years in the Cuyo Andean Region of Argentina', *Intern. J. Climatol.* **9**, 55–68.
Pittock, A. B.: 1980, 'Patterns of Climatic Variation in Argentina and Chile. I. Precipitation, 1931–1960', *Mon. Wea. Rev.* **108**, 1347–1361.
Politis, G. C.: 1984, 'Climatic Variations during Historical Times in Eastern Buenos Aires Pampas, Argentina', *Quatern. South Amer. Antarc. Penins.* **2**, 133-162.

Quinn, W. H. and Neal, V. T.: 1992, 'The Historical Record of El Niño events', in Bradley, R. S. and Jones, P. D. (eds.), *Climate since A. D. 1500*, Routledge, London, pp. 623–648.

Quinn, W. H., Neal, V. T., and Antunez de mayolo, S. E.: 1987, 'El Niño Occurrences over the Past Four and a Half Centuries', *J. Geophys. Res.* **92**, 14449–14461.

Rabassa, J. and Clapperton, C. M.: 1990, 'Quaternary Glaciations of the Southern Andes', *Quatern. Sci. Rev.* **2**, 153–174.

Ropelewski, C. F. and Halpert, M. S.: 1987, 'Global and Regional Scale Precipitation Patterns Associated with the El Niño/Southern Oscillation', *Mon. Wea. Rev.* **115**, 1606–1626.

Röthlisberger, F.: 1986, *1000 Jahre Glestschergerchichte der Erde*, Salzburg: Verlag Sauerlander.

Rubin, M. J.: 1955, 'An Analysis of Pressure Anomalies in the Southern Hemisphere', *Notos* **4**, 11–16.

Rutlland, J., and Fuenzalida, H.: 1991, 'Synoptic aspects of the central Chile rainfall variability associated with the Southern Oscillation', *International Journal of Climatology* **11**, 63–76.

Stuiver, M. and Becker, B.: 1986, 'High-Precision Decadal Calibration of the Radiocarbon Time Scale, A.D. 1950–2500 BC', *Radiocarbon* **28**, 863–910.

Taljard, J. J.: 1972, 'Synoptic Meteorology of the Southern Hemisphere', *Meteorol. Monogr.* **13**, 139–213.

Taulis, E.: 1934, 'De la Distribution des Pluies au Chile', *Materoux pour l'Etude des Calamites*, Part 1, Societe de Geographie de Geneve, pp. 3–20.

Van der Hammen, T.: 1991, 'Palaeoecological Background: Neotropics', *Clim. Change* **19**, 37–47.

Villalba, R.: 1990a, 'Climatic Fluctuations in Northern Patagonia in the Last 1000 Years as Inferred from Tree-Ring Records', *Quatern. Res.* **34**, 346–360.

Villalba, R.: 1990b, 'Latitude of the Surface High-Pressure Belt over Western South America during the Last 500 Years as Inferred from Tree-Ring Analysis', *Quatern. South Amer. Antarct. Penins.* **7**, 273–303.

Villalba, R., Leiva, J. C., Rubulis, S., Suarez, J. A., and Lenzano, L.: 1990, 'Climate, Tree-Ring and Glacial Fluctuations in the Río Frías Valley, Río Negro, Argentina', *Arc. Alp. Res.* **22**, 215–232.

Williams, L. D. and Wigley, T. M. L.: 1983, 'Comparison of Evidence for Late Holocene Summer Temperature Variations in the Northern Hemisphere', *Quatern. Res.* **20**, 286–307.

Wells, L. E.: 1987, 'An Alluvial Record of El Niño Events from Northern Coastal Perú', *J. Geophys. Res.* **92**, 14463–14470.

Wells, L. E.: 1990, 'Holocene History of the El Niño Phenomenon as Recorded in Flood Sediments of Northern Coastal Perú', *Geology* **18**, 1134–1137.

(Received 22 September, 1992; in revised form 27 October, 1993).

TREE-RING RECONSTRUCTED RAINFALL OVER THE SOUTHEASTERN U.S.A. DURING THE MEDIEVAL WARM PERIOD AND LITTLE ICE AGE

DAVID W. STAHLE   and   MALCOLM K. CLEAVELAND

Tree-Ring Laboratory, University of Arkansas, Fayetteville, AR 72701, U.S.A.

Abstract. A 1053-year reconstruction of spring rainfall (March-June) was developed for the southeastern United States, based on three tree-ring reconstructions of statewide rainfall from North Carolina, South Carolina, and Georgia. This regional reconstruction is highly correlated with the instrumental record of spring rainfall ($r = +0.80$; 1887–1982), and accurately reproduces the decade-scale departures in spring rainfall amount and variance witnessed over the Southeast during the past century. No large-magnitude centuries-long trends in spring rainfall amounts were reconstructed over the past 1053 years, but large changes in the interannual variability of spring rainfall were reconstructed during portions of the Medieval Warm Period (MWP), Little Ice Age (LIA), and the 20th century. Dry conditions persisted at the end of the 12th century, but appear to have been exceeded by a reconstructed drought in the mid-18th century. High interannual variability, including five extremely wet years were reconstructed for a 20-yr period during the late 16th and early 17th centuries, and may reflect amplified atmospheric circulation over eastern North America during what appears to have been one of the most widespread cold episodes of the Little Ice Age.

1. Introduction

The Medieval Warm Period (MWP) and Little Ice Age (LIA) have received increased scrutiny as possible examples of centuries-long, global-scale climate anomalies caused by natural factors internal and/or external to the climate system (e.g., Folland *et al.*, 1990; Bradley and Jones, 1992). The true scope and magnitude of these late Holocene climate episodes have become important issues relevant to the detection of anthropogenic influences on global climate. For example, some fraction of the observed rise in surface air temperature during the 20th century may represent a natural amelioration of the LIA (Folland *et al.*, 1990). Also, the regional climate and ecological changes during the MWP might provide useful examples of the environmental impacts that could attend CO_2 induced global warming. It is therefore important to document the nature of these climate episodes in all regions and all four seasons using as wide a range of climate variables as possible.

The MWP and LIA have been most clearly identified in a variety of historical and proxy climate sources from the temperate and arctic latitudes of western Europe and North America (e.g., LaMarche, 1974; Le Roy Ladurie, 1971; Williams and Wigley, 1983; Karlen, 1988; Chapman and Clow, 1991; Briffa *et al.*, 1992). The MWP dates from approximately A.D. 1000 to 1300 and included retreating glaciers (Grove, 1988), above average growth in temperature sensitive trees of the western

Climatic Change **26**: 199–212, 1994.
© 1994 *Kluwer Academic Publishers.*

United States and Scandinavia (LaMarche, 1974; Briffa *et al.*, 1992), and Norse settlement of Greenland (Lamb, 1977). The LIA dates from approximately A.D. 1550 to 1850 and included roughly synchronous glacial advances in Europe and North America (e.g., Williams and Wigley, 1983; Grove, 1988; Bradley and Jones, 1992), and a series of long, cold winters in western Europe (Lamb, 1977). Evidence for the larger, perhaps global-scale extent of these late Holocene climatic episodes has been reported from Russia (Graybill and Shiyatov, 1992), China (Fu, 1990), Tasmania (Cook *et al.*, 1991), and South America (e.g., Thompson *et al.*, 1986; Villalba, 1990). However, scrutiny of a diverse body of climate evidence covering the last 500 yr led Jones and Bradley (1992) to conclude that the nature and timing of regional climate anomalies was quite variable during this period, and the consistent worldwide climatic conditions suggested by the term Little Ice Age might misrepresent the data. Examination of the proxy data reviewed by Williams and Wigley (1983) suggests that regional variability in reconstructed climates may cast similar doubts over the general use of the term Medieval Warm Period. There certainly has been considerable regional variability in seasonal and annually-averaged temperatures during the past century when hemispheric and globally-averaged temperature data have indicated a warming trend of approximately 0.5°C (e.g., Folland *et al.*, 1990).

Some of the clearest evidence for the LIA and MWP has been derived from glaciological and pollen data (e.g., Grove, 1988), but the temporal resolution and specific climatological implications of this evidence are often poorly constrained. Much of the evidence for the LIA and MWP is also based on proxies which are believed to be largely sensitive to variations in surface air temperature during the summer or winter season. Accurate high resolution paleoclimatic estimates of precipitation are generally less common, so the possible change in regional to global scale hydrological conditions during the MWP and LIA are not well documented, particularly during the transitional seasons of spring and fall. Dramatic, post-glacial changes in regional hydrology have occurred in regions such as North Africa (Folland *et al.*, 1990) and possibly over the southeastern U.S.A. (e.g., Frey, 1954; Whitehead, 1972), particularly during the mid-Holocene thermal maximum ca. 6000 B.P. The possibility that significant hydrological changes occurred during the MWP and LIA is therefore an important question, and can be addressed in some regions like the southeastern United States with high resolution tree-ring chronologies. In this paper, we develop a 1053-yr reconstruction of spring rainfall over the southeastern U.S.A., and then search for interannual to decadal anomalies in spring rainfall during the MWP, LIA, and other episodes over the past millennium.

2. Reconstruction of Spring Rainfall

Five rainfall-sensitive tree-ring chronologies 780 to 1614-yr long have been developed from baldcypress (*Taxodium distichum*) trees in North Carolina, South Carolina, and Georgia (Figure 1). These chronologies are well correlated with each

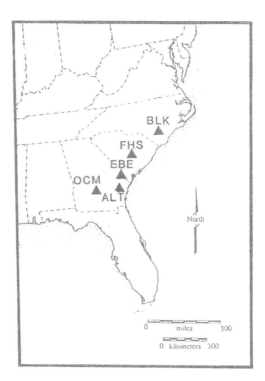

Fig. 1. The locations of the five tree-ring chronologies (triangles) used to develop the spring rainfall reconstruction for the 'Southeast' (i.e., averaged over North Carolina, South Carolina, and Georgia). BLK = Black River; FHS = Four Holes Swamp; EBE = Ebenezer Creek; ALT = Altamaha River; OCM = Ocmulgee River.

other and with monthly precipitation totals from March though June averaged on a statewide basis by Karl *et al.* (1983a, b, c). Variations in water level and water quality (particularly the dissolved oxygen concentration), which are both linked with rainfall amounts, are believed to be involved in the baldcypress growth response to spring rainfall (Stahle and Cleaveland, 1992). The radial growth of baldcypress is also inversely correlated with percentage possible sunshine and temperature during the growing season, but these influences are quite weak compared to the rainfall signal (e.g., Stahle *et al.*, 1991).

    In a previous study, the five baldcypress chronologies were used to reconstruct spring rainfall amounts in each of North Carolina (April-June), South Carolina, and Georgia (both March to June; Stahle and Cleaveland, 1992). All five chronologies were considered for the calibration of spring rainfall in each state, and the best subsets of chronologies were selected (based on the *F*-statistics and explained variance computed in stepwise multiple regression analyses). The Black River chronology alone was used in a bivariate regression model to reconstruct April through June rainfall in North Carolina; the Four Holes Swamp and Ebenezer Creek chronologies were used in a multiple regression model to reconstruct March through June rainfall in South Carolina; and a regional average of the Altamaha and Ocmulgee River chronologies was used in bivariate regression to calibrate and reconstruct March through June rainfall in Georgia (Stahle and Cleaveland, 1992). Because the Ocmulgee chronology starts in A.D. 1206, the Altamaha-Ocmulgee average chronology is based only on Altamaha prior to 1206, and the variance of Altamaha before 1206 was reduced to match the variance of the Altamaha-Ocmulgee combination (1206–1985). The selected tree-ring predictor(s) explained highly significant fractions of the observed spring rainfall variance in each state (i.e., R^2 = 0.54; 0.58, and 0.68 based on the period 1887 (or 1892 in Georgia) to 1936 in North Carolina, South Carolina, and Georgia, respectively. The reconstructions of state-averaged rainfall derived from the tree-ring data in each state were also well verified when compared with instrumental state-averaged rainfall data available from 1937 to 1982 (Karl *et al.*, 1983 a, b, c), which were withheld from the calibration (Stahle and Cleaveland, 1992).

The three state ('Southeast') precipitation average in this study uses the three independent reconstructions of state-average precipitation for North and South Carolina and Georgia (Stahle and Cleaveland, 1992), all very well correlated over their 981-yr common period. We optimized the North Carolina tree-ring – climate relationship by reconstructing April-June rainfall, but March-June precipitation correlated better with growth in South Carolina and Georgia. All three reconstructions were square root transformed to adjust non-normal distributions. Because the three rainfall reconstructions covered two slightly different periods and had different means and variances, we normalized all three series before averaging (transformed to mean = 0.0, variance = 1.0), then normalized the resulting average from A.D. 933 to 1985 (Figures 2, 3, and 4).

This southeastern proxy rainfall series has uneven sample size through time, and its reliability certainly decreases with the absence of the Ocmulgee data before A.D. 1206 and especially with the absence of the South Carolina data before A.D. 1005. Nevertheless, the reconstruction from 1005 to 1206 is based on from 41 to 89 exactly dated ring-width series from four sites (representing 21 to 30 separate trees), and should represent major episodes of drought and wetness during this important part of the MWP with reasonable accuracy. The southeastern rainfall reconstruction during the LIA (A.D. 1600) is very well replicated, with 161 ring-width series from 97 separate trees at all five collection sites.

To test the accuracy of the southeastern spring rainfall reconstruction, we compared the reconstructed data with instrumental rainfall data during the period from 1887 to 1982 (Figure 2). The instrumental spring rainfall average for the 'Southeast' was developed in the same manner as the reconstructed average. April to June total rainfall in North Carolina, and March to June total rainfall in South Carolina and Georgia were first individually normalized, averaged together on an annual basis from 1887 to 1982, and the resulting average was then normalized. Note that the instrumental rainfall data for North and South Carolina both begin in 1887, but the data for Georgia begins in 1892 (Karl *et al.*, 1983a, b, c). Consequently, the

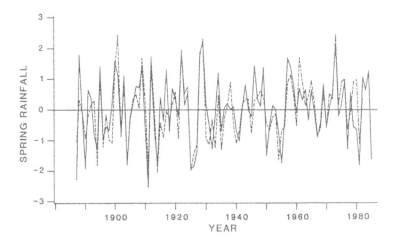

Fig. 2. Observed (dashed line) and reconstructed (solid line) spring rainfall (March-June) averaged and normalized for the southeastern United States from 1887 to 1982.

observed Southeast average is based on only two states from 1887 to 1891.

The reconstructed southeastern rainfall data are significantly correlated with the instrumental average at both the annual and decadal timescales (Figures 2 and 3). The Pearson correlation coefficient (r) computed for the full period of common data from 1887–1982 was +0.80 ($P < 0.0001$). However, the relationship between reconstructed and observed rainfall was higher during the first half of the common period (i.e., $r = +0.84$ for 1887 to 1936, while $r = +0.71$ for 1937 to 1982, see Figure 2). The reason for the lower correlation after 1936 is not entirely clear, but the period from 1887 to 1936 was characterized by higher variance in the tree-ring and rainfall data (e.g., Figures 5, 6), and the spatial coherence of spring rainfall departures over the Carolinas and Georgia appear to have declined after 1936 (Stahle and Cleaveland, 1992). There are also differences in the treatment of the instrumental rainfall data used to develop the state averages before and after 1931 (Karl *et al.*, 1983a, b, c), which are discussed further below. Nevertheless, the reconstructed rainfall data are strongly correlated with the instrumental data in both subperiods and overall, and these correlation results suggest that the reconstruction represents some 50 to 70% of the actual rainfall variance during 30 to 40-yr subperiods back to A.D. 1206, and somewhat less back to A.D. 1005.

The decade-scale fluctuations evident in the instrumental rainfall data are well replicated by the reconstructed data (Figure 2). This low-frequency coherence is illustrated in Figure 3 where the instrumental and reconstructed spring rainfall data from 1887 to 1982 were both filtered with a smoothing spline designed to emphasize variance in the 10-yr range. The decade-scale variations in observed and reconstructed rainfall data (Figure 3) represent 13.6 and 10.3% of the variance in the annual time series of spring rainfall from 1887 to 1982 (Figure 2), respectively.

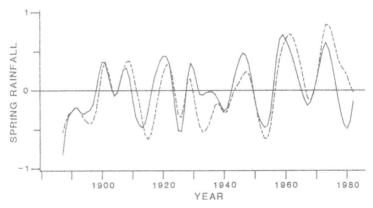

Fig. 3. The observed (dashed line) and reconstructed (solid line) spring rainfall for the three-state southeastern average (1887–1982) filtered with a smoothing spline designed to reduce 50% of the variance in a sine wave with a frequency of 10 yr (Cook and Peters, 1981). Note the agreement of decadal excursions between the observed and reconstructed series.

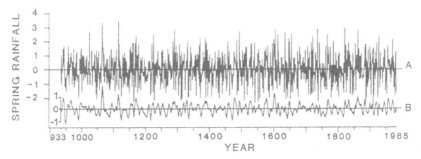

Fig. 4. (A) The normalized tree-ring reconstruction of southeastern spring rainfall (March-June) averaged over the Carolinas and Georgia from A.D. 933 to 1985; (B) The spline filtered version of the spring rainfall reconstruction emphasizing variations in the 10-yr frequency range.

The decadal variations in the reconstruction replicate the timing and magnitude of the low frequency variations in spring rainfall with good accuracy in most cases. The general upward trend instrumental in spring rainfall previously reported for the Carolinas and Georgia by Stahle and Cleaveland (1992) is reproduced by this tree-ring reconstruction, which suggests that the trend is not an artifact of precipitation measurement or data processing.

3. Analysis of Spring Rainfall during the Past Millennium

The spring rainfall reconstruction for the Southeast is plotted annually from A.D. 933 to 1985 in Figure 4a. This reconstruction is dominated by high frequency variance which is typical of the instrumental spring rainfall data available for the past century (Figure 2). There are no prominent century-scale trends in reconstructed spring rainfall which might reflect the regional influence of the Medieval

Warm Period or Little Ice Age in the southeastern United States. However, without specialized techniques (e.g., Cook *et al.*, 1990; Briffa *et al.*, 1992), tree-ring chronologies in general, and baldcypress chronologies in particular, are not ideal for detecting very long century-scale climate trends. The general necessity for removing age-related growth trends in tree-ring data limits the low-frequency climate signal in these data. The unique wetland adaptation of baldcypress further complicates the recovery of century-scale climate signal.

Baldcypress growth appears to be particularly sensitive to dissolved oxygen levels in swamp waters, and the fine root system of these trees tends to stratify in the well-oxygenated near surface waters (Stahle and Cleaveland, 1992). If the mean water level were to change permanently, baldcypress can, within limits, sprout new root hairs to follow persistent changes in the zone of well-oxygenated water. This root habit of baldcypress was dramatically illustrated at Reelfoot Lake, Tennessee, following the cataclysmic earthquakes of 1811–1812. These great earthquakes permanently raised the water level some 2 m to form Reelfoot Lake, and the baldcypress which survived responded with a phenomenal acceleration in growth which lasted for at least 10 yr. However, these trees also eventually formed a 'hanging buttress' and fine root system in the near surface waters, presumably to exploit the higher concentrations of dissolved oxygen (Stahle *et al.*, 1992). This vertical migration of the fine root system to exploit water level changes acts as a natural high-pass filter and appears capable of attenuating the registration of long-term rainfall trends in baldcypress tree-ring chronologies. The time involved in this physiological response has not been carefully quantified, but appears to have taken at least 30 yr following the formation of Reelfoot Lake. The apparent ability of baldcypress to adapt to long-term water-level changes is certainly antagonistic to the registration of possible century-scale rainfall trends and may diminish as well the registration of multi-decadal rainfall changes in baldcypress chronologies. Nevertheless, Figures 2 and 3 demonstrate that baldcypress ring-width chronologies are excellent proxies of interannual to decadal-scale climate variance.

The spring rainfall reconstruction was smoothed with a cubic spline to emphasize variance in the 10-yr range in Figure 4b, and the smoothed reconstruction indicates that the tendency for spring rainfall to oscillate between decades of relative drought and wetness as observed during the 20th century (Figure 3), has been prevalent over the past 1000 yr. The decadal variability in Figure 4b represents 12.3% of the variance in the rainfall reconstruction which is similar to the ratio represented by the 10-yr filter of the instrumental rainfall data. These decade-scale excursions in spring rainfall have important socioeconomic and environmental implications, but the frequency of wet or dry decades does not appear to have changed dramatically during the MWP, LIA, or at any other time during the past millennium.

To further examine spring rainfall variations over the past millennium and test for statistically significant changes in rainfall amounts or variance during the MWP, LIA, and other episodes, the rainfall reconstruction was arbitrarily subdi-

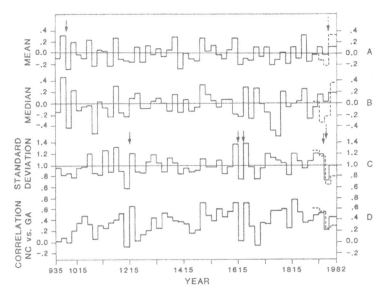

Fig. 5. The mean (A); median (B); and standard deviation (C) of spring rainfall over the southeastern United States computed for non-overlapping 20-yr subperiods from A.D. 935 to 1982 (solid lines). Statistically significant differences between adjacent 20-yr subperiods are indicated by arrows ($P \leq 0.05$). These statistics were also computed for the instrumental spring rainfall data during the same subperiods since 1895 (dashed line). The last subperiod includes 28 yr (i.e., 1955–1982); The correlation (D) between spring rainfall for North Carolina and Georgia for 20-yr subperiods [reconstructed series (solid line) from Stahle and Cleaveland (1992); instrumental data (dashed line)].

vided into non-overlapping 20-yr periods from A.D. 935 to 1982 (i.e., 935–954, 955–974...1935–1954, 1955–1982). For comparative purposes, the instrumental spring rainfall data for the Southeast were also subdivided into the same 20-yr subperiods from 1895–1982. In both cases, the last subperiod (1955–1982) included 28 yr (1955–1982). This testing scheme is certainly not exhaustive, but should be conservative. The arbitrary starting time (A.D. 935) and test interval length (20 yr) could fail to bracket the particular timing of a large excursion in the mean or variance, but it is unlikely that many significant changes in spring rainfall amount or variance would be detected if large excursions were not in the data. We only discuss the test results based on the 20-yr subperiods, but note that these analyses were repeated using 25, 30, and 35-yr subperiods. The results obtained with the longer subperiods closely resembled the 20-yr test interval results discussed below.

The mean, median, and standard deviation were computed for each 20-yr subperiod (Figure 5a, b, c), and differences between adjacent subperiods were tested for significance using t-tests, a non-parametric test between medians (SAS Institute, 1989), and Bartlett's test (Steele and Torrie, 1980), respectively. The time series plot of 20-yr means (Figure 5a) indicates that reconstructed spring rainfall was below average during much of the MWP (i.e., 9 of 15 subperiods from 975–1274), and was above average during portions of the LIA (i.e., 5 of 7 subperiods from 1475-1614). However, the spring rainfall means computed for each of these long time intervals are not statistically different from each other or from the long-term mean. Furthermore, when the means of adjacent 20-yr subperiods were tested for differences, only the change from 975–994 to 995–1014 was significant (Figure 5a). These data suggest that the MWP and LIA did not involve large-magnitude changes in spring rainfall amounts over the southeastern United States. However, the significant change in instrumentally-recorded rainfall from 1935–1954 to 1955–1982 ($P \leq 0.05$, Figure 5a) was not fully reproduced by the tree-ring reconstruction. This is further evidence that the tree-ring data are underestimating the magnitude and possibly the statistical significance of spring rainfall fluctuations over the past millennium.

The plot of median spring rainfall for the 20-yr subperiods largely conform with the 20-yr means, with the exception of the mid-18th century when 60 yr of below median rainfall were reconstructed. This 18th century drought era is evident in the smoothed reconstruction as the most prolonged period of below normal spring rainfall in the past 1000 yr (Figure 4b). This drought appears to have equalled or exceeded any dry episodes of comparable length during the MWP, and indicates that the time period normally associated with the LIA was certainly not uniformly wet (or dry) during spring and early summer over the southeastern United States.

Although no highly significant century-scale changes in spring rainfall amount have been reconstructed over the past 1000 years, there do appear to have been large changes in spring rainfall variance during the Medieval Warm Period, the 20th century, and most notably during the Little Ice Age (Figure 5c). Bartlett's test indicates that the standard deviation of reconstructed rainfall was significantly different ($P \leq 0.05$) between the following pairs of 20-yr subperiods: 1195–1214 and 1215–1234, 1595–1614 and 1615–1634, 1615–1634 and 1635–1654, and 1915–1934 and 1935–1954 (Figure 5c). Because 52 different 20-yr subperiods were tested, unequal variance might be observed between 2.6 subperiods simply by chance (assuming $P \leq 0.05$). The four periods of unequal variance may therefore suggest that spring rainfall is subject to statistically significant change over the southeastern United States.

The significant decline in spring rainfall variance observed in the instrumental record from 1915–1934 to 1935–1955 was matched by the reconstruction, and indicates that the tree-ring data are accurately reproducing decade-scale changes in rainfall variability. The tree-ring data also provide some reassurance that the variance changes evident in the instrumental rainfall data for the Southeast before and after 1931 are not artifacts of rainfall measurement or computation. The monthly precipitation data for the three states used to compile the Southeast average were based on areally-weighted climatic division data after 1931, but on equally-weighted station data prior to 1931 that were adjusted to 'more closely resemble' the state averages based on the divisional data (Karl *et al.*, 1983a, b, c). The faithful reproduction of the instrumentally-recorded changes in rainfall variance before and after 1931 by the tree-ring reconstruction (Figure 2), and by the three individual spring rainfall reconstructions for North Carolina, South Carolina, and Georgia (Stahle and Cleaveland, 1992) suggest that the variance changes in observed spring rainfall over the Southeast during the past century were largely real.

Spring rainfall variance peaked in the 17th century, which may represent the influence of unusual atmospheric circulation over the Southeast and much of the Northern Hemisphere during the Little Ice Age. Jones and Bradley (1992) suggest that the period from 1590 to 1610 was one of the few periods during the Little Ice Age when the paleoclimatic data they reviewed indicated synchronous cool conditions on a hemispheric to global scale. Southeastern spring rainfall is estimated to have been above average from 1595 to 1614, but also highly variable and with more dry years than wet years (Figures 5a, b, c). In fact, 5 of the 48 wettest years over the past millennium (i.e., in the upper 5th percentile) were reconstructed during this 20 year interval (1596, 1600, 1602, 1605, and 1613), which is over five times higher than the rate at which these wet extremes would occur if they were randomly distributed through time.

The interannual variability of reconstructed spring rainfall was also elevated during four 20-yr subperiods from the 11th to 13th centuries, and the increase from 1195–1214 to 1215–1234 was significant (Figure 5c). Whether these variance changes represent an effect of the Medieval Warm Period on spring rainfall over the Southeast is less clear due to the decline in sample size prior to A.D. 1206 associated with the loss of the Ocmulgee chronology.

The possibility that periods of high rainfall variability may reflect an amplified atmospheric circulation can be checked during the 20th century when both the observed and reconstructed rainfall data for the Southeast exhibit a significant change in variance. Spring rainfall declined considerably after the 1930's (Figure 5c), so the rainfall data were split into two long subperiods for this analysis (i.e., 1895–1930 and 1931–1982). The decline in rainfall variance between these two long subperiods is also significant for both the instrumental and reconstructed data ($P < 0.01$ and 0.05, respectively). Regional sea level pressure data and a circulation index related to the North Atlantic subtropical high were then used to investigate possible circulation changes between these two subperiods (Table I).

Sea level pressure data available from 1899 to 1980 were interpolated from station data to a $10° \times 10°$ latitude/longitude grid (Trenberth and Paolino, 1980), and we used the regional data for March through June at 30°N/90°W. Observed and reconstructed spring rainfall for the Southeast are significantly correlated with this regional sea level pressure series from 1899 to 1980 (Table I). However, this relationship is considerably stronger prior to 1931 (Table I), when the standard deviation of observed and reconstructed rainfall was highest (Figure 5c).

Anomalies in the zonal position of the North Atlantic subtropical high have been linked with observed and reconstructed spring rainfall over the Carolinas and Georgia (Stahle and Cleaveland, 1992). An index of the zonal movement of the

TABLE I: Correlation analyses between spring rainfall in the southeastern United States and atmospheric circulation data

	Regional Sea Level Pressure (grid point at 30° N–90° W)		
	1899–1930	1931–1980	1899–1980
Spring Rainfall			
Observed	−0.57***	−0.31**	−0.41***
Reconstructed	−0.52***	−0.27*	−0.35***
	Extended North Atlantic Ridge Index (June)		
	1895–1930	1931–1980	1895–1980
Spring Rainfall			
Observed	−0.31*	−0.24*	−0.26**
Reconstructed	−0.40**	−0.01	−0.19*

* $= P \leq 0.10$
** $= P \leq 0.05$
*** $= P \leq 0.01$

subtropical high into the eastern United States during summer has recently been developed by Heim *et al.* (1993), based on spatial anomalies in temperature and precipitation data. This Extended North Atlantic Ridge Index (or Ridge Index) is not available for spring, but the Ridge Index for June is nevertheless significantly correlated with observed and reconstructed rainfall totals for the spring season (i.e., March through June, Table I). This correlation between southeastern rainfall and the June Ridge Index is also highest during the early subperiod (1899–1930), particularly for the reconstructed data (Table I).

These correlation results suggest that spring rainfall over the Southeast was more strongly influenced by large-scale circulation during the early 20th century period of high rainfall variability, and this may also have been true during periods of high rainfall variability reconstructed for portions of the LIA and MWP (Figure 5c). These inferences are supported in part by analyses of the spatial homogeneity of spring rainfall over the Southeast, which appears to be highest during periods of greatest circulation influence. High interannual variability in the regional rainfall data usually reflects strong agreement among the three statewide rainfall series used to compute the regional average. Poor agreement among the three statewide series would tend to cancel individual state departures and thus reduce variance in the derived regional average. This is illustrated in Figure 5d by correlation analyses between the statewide spring rainfall data for North Carolina and Georgia, computed for the same 20-yr subperiods from A.D. 935 to 1982 for the reconstructed series (Stahle and Cleaveland, 1992), and from 1895 to 1982 for the instrumental series (Karl *et al.*, 1983a, b). The periods of strongest correlation between North Carolina and Georgia rainfall are usually also periods of highest variance in the regional rainfall average (Figure 5c, d).

The spatial homogeneity of observed and reconstructed spring rainfall from North Carolina to Gerogia was highest prior to the 1930's when regional sea level pressure and the Ridge Index were most strongly correlated with the southeastern spring rainfall data (Figure 5d, Table I). The decline in the correlation between North Carolina and Georgia rainfall after the 1930's (Figure 5d) is also evident in spring rainfall data for 11 individual weather stations located throughout the Carolinas and Georgia (Stahle and Cleaveland, 1992). One explanation for this apparent decline in the spatial homogeneity of spring rainfall over the Southeast after the 1930's might be a shift in the frequency of frontal as opposed to convective precipitation associated with a decrease in the influence of large-scale circulation during spring. These possibilities are important to the interpretation of reconstructed rainfall and could be tested with more detailed analyses of rainfall and circulation data for the Southeast.

4. Conclusions

The reconstructions of spring rainfall using baldcypress tree-ring chronologies indicate that the Medieval Warm Period and Little Ice Age were not vividly reflected in prolonged rainfall deficits or surpluses over the Carolinas and Georgia. There is weak evidence that dry conditions were more prevalent from ca. 1040 to 1275, and that wet conditions were prevalent from 1475 to 1620, but these differences are small and are not statistically significant. In fact, the most prolonged dry episode was reconstructed during the mid-18th century, at a time usually associated with the LIA. As presently formulated, these baldcypress tree-ring chronologies are not well suited for detecting possible century-scale excursions in spring rainfall amounts, but they do appear to be faithfully recording decade-scale changes. These reconstructed decade-scale variations in spring rainfall amount do not clearly identify MWP or LIA climate episodes over the Southeast.

The reconstructed data do, however, suggest that spring rainfall is subject to significant decadal changes in variance over the southeastern United States. Large and significant changes in spring rainfall variance were reconstructed during the 13th, 17th, and 20th centuries, and may represent the influence of unusual atmospheric circulation conditions during the MWP and LIA over the Southeast. The period from A.D. 1595 to 1614 is especially interesting because it appears to have been one of the most consistent and extensive cold episodes of the Little Ice Age, and the high mean, low median, and high variance of reconstructed rainfall indicate that a few extremely wet years were influential in the statistical properties of spring rainfall. These rainfall extremes suggest that large-scale atmospheric circulation over eastern North America may have been amplified during these particular years, and was perhaps responsible for climate anomalies elsewhere in the Northern Hemisphere during this notable episode of the Little Ice Age. Southeastern spring rainfall was more highly correlated with large-scale circulation indices during a period of high variability in the early 20th century, which lends some credence to these circulation inferences.

Acknowledgements

The research was supported by the National Science Foundation, Climate Dynamics Program, under grant number ATM-8914561. We thank Dr. Charles Wharton, Julie Moore, and Norman Brunswig for assistance in locating the old growth baldcypress stands sampled for this analysis, Dr. H.C. Fritts and G.R. Lofgren of the University of Arizona Laboratory of Tree-Ring Research for providing the gridded sea level pressure data, and Dr. R.R. Heim, Jr. for the Extended North Atlantic Ridge indices. We thank the landowners, particularly the National Audubon Society, the Nature Conservancy, Mr. Nelson Squires, and the Benjamin Cone estate, for permission to non-destructively sample the growth rings from these ancient forests.

References

Bradley, R. S. and Jones, P. D.: 1992, *Climate since A.D. 1500*, Routledge, London.
Briffa, K. R., Jones, P. D., Bartholin, T. S., Eckstein, D., Schweingruber, F. H., Karlen, W., Zetterberg, P., and Eronen, M.: 1992, 'Fennoscandian Summers from A.D. 500: Temperature Changes on Short and Long Timescale', *Clim. Dynam.* **7**, 111–119.
Chapman, D. S. and Clow, G. D.: 1991, 'Surface Temperature Histories Reconstructed from Borehole Temperatures: A Geothermal Contribution to the Study of Climate Change', *EOS* supplement. Oct. 29, 1991, pp. 69.
Cook, E. R. and Peters, K.: 1981, 'The Smoothing Spline: A New Approach to Standardizing Forest Interior Ring-Width Series for Dendroclimatic Studies', *Tree-Ring Bull* **41**, 45–53.
Cook, E. R., Briffa, K. R., Shiyatov, S. G., and Mazepa, V.: 1990, 'Tree-Ring Standardization and Growth Trends Estimation', in Cook, E. R. and Kairiukstis, L. A. (eds.), *Methods of Dendrochronology*, Kluwer, Dordrecht.
Cook, E. R., Bird, T., Peterson, M., Barbetti, M., Buckley, B., D'Arrigo, R., Francey, R., and Tans, P.: 1991, 'Climatic Change in Tasmania Inferred from a 1089-Year Tree-Ring Chronology of Huon Pine', *Science* **253**, 1266–1268.
Folland, C. K., Karl, T. R., and Vannikov, K. Ya.: 1990, 'Observed Climate Variations and Change', in Houghton, J. T., Jenkins, G. J., and Ephraums, J. J. (eds.), *Climate Change; the IPCC Scientific Assessment*, Cambridge University Press, pp. 194–238.
Frey, D. G.: 1954, 'Evidence for the Recent Enlargement of the "Bay" Lakes of North Carolina', *Ecology* **35**, 78–88.
Fu Congbin: 1990, 'Climatic Change in China from the Preinstrumental Period to the Last 100 Years', in Parker, D. W. (ed.), *Observed Climate Variations and Change: Contributions in Support of Section 7 of the 1990 IPCC Scientific Assessment*, Intergovernmental Panel on Climate Change, Geneva.
Graybill, D. A. and Shiyatov, S. G.: 1992, 'Dendroclimatic Evidence from the Northern Soviet Union', in Bradley, R. S. and Jones, P. D. (eds.), *Climate since A.D. 1500*, Routledge, London, pp. 393–414.
Grove, J. M.: 1988, *The Little Ice Age*, Methuen, London.
Heim, R. R., Jr., Brown, W. O., Owenby, J. R., and Garvin, C.: 1993, 'Circulation Indices for the Contiguous United States: 1895–1992', *Proceedings of the 17th Annual Climate Diagnostics Workshop*, U.S. Dept. of Commerce, Washington D.C.
Jones, P. D. and Bradley, R. S.: 1992, 'Climatic Variations over the Last 500 Years', in Bradley, R. S. and Jones, P. D. (eds.), *Climate since A.D. 1500*, Routledge, London, pp. 649–666.

Karl, T. R., Metcalf, L. K., Nicodemus, M. L., and Quayle, R. G.: 1983a, 'Statewide Average Climatic History, Georgia, 1892–1982', *Historical Climatology Series 6–1*, National Climatic Data Center, Asheville, North Carolina.

Karl, T. R., Metcalf, L. K., Nicodemus, M. L., and Quayle, R. G.: 1983b, 'Statewide Average Climatic History, North Carolina, 1887–1982', *Historical Climatology Series 6–1*, National Climatic Data Center, Asheville, North Carolina.

Karl, T. R., Metcalf, L. K., Nicodemus, M. L., and Quayle, R. G.: 1983c, 'Statewide Average Climatic History, South Carolina, 1887–1982', *Historical Climatology Series 6–1*, National Climatic Data Center, Asheville, North Carolina.

Karlen, W.: 1988, 'Scandinavian Glacial and Climatic Fluctuations during the Holocene', *Quatern. Sci. Rev.* **7**, 199–209.

LaMarche, V. C., Jr.: 1974, 'Paleoclimatic Inferences from Long Tree-Ring Records', *Science* **183**, 1043–1048.

Lamb, H. H.: 1977, *Climate: Present, Past, and Future*, Volume 2, Methuen, London.

Le Roy Ladurie, E.: 1971, *Times of Feast, Times of Famine*, Doubleday, New York.

SAS Institute, Inc.: 1989, *SAS/STAT User's Guide*, Version 6, 4th edition, Vol. 2, SAS Institute, Inc., Cary, North Carolina.

Stahle, D. W., Cleaveland, M. K., and Cerveny, R. S.: 1991, 'Tree-Ring Reconstructed Sunshine Duration over Central USA', *Int. J. Climatol.* **11**, 285–295.

Stahle, D. W. and Cleaveland, M. K.: 1992, 'Reconstruction and Analysis of Spring Rainfall over the Southeastern U.S.A. for the Past 1000 Years', *Bull. Amer. Meteorol. Soc.* **73**, 1947–1961.

Stahle, D. W., Van Arsdale, R. B., and Cleaveland, M. K.: 1992, 'Tectonic Signal in Baldcypress Trees at Reelfoot Lake, Tennessee', *Seismol. Res. Letters* **63**, 439–447.

Steele, R. G. D. and Torrie, J. H.: 1980, *Principles and Procedures of Statistics, Second Edition*, McGraw-Hill, New York.

Thompson, L. G., Mosley-Thompson, E., Dansgaard, W., and Grootes, P. M.: 1986, 'The Little Ice Age as Recorded in the Stratigraphy of the Tropical Quelccaya Ice Cap', *Science* **234**, 361–364.

Trenberth, K. E. and Paolino, D. A.: 1980, 'The Northern Hemisphere Sea Level Pressure Data Set: Trends, Error and Discontinuities', *Mon. Wea. Rev.* **108**, 855–872.

Villalba, R.: 1990, 'Climatic Fluctuations in Northern Patagonia during the Last 1000 years as Inferred from Tree-Ring Records', *Quatern. Res.* **34**, 346–360.

Whitehead, D. R.: 1972, 'Developmental and Environmental History of the Dismal Swamp', *Ecol. Monogr.* **42**, 301–315.

Williams, L. D. and Wigley, T. M. L.: 1983, 'A Comparison of Evidence for Late Holocene Summer Temperature Variations in the Northern Hemisphere', *Quatern. Res.* **20**, 283–307.

(Received 22 September, 1992; in revised form 18 October, 1993)

MIDDLE AGES TEMPERATURE RECONSTRUCTIONS IN EUROPE, A FOCUS ON NORTHEASTERN ITALY*

FRANÇOISE SERRE-BACHET[†]

URA CNRS 1152, Laboratoire de Botanique historique et Palynologie, Faculté des Sciences et Techniques de St-Jérôme, 13397 Marseille cedex 13, France

Abstract. In the set of climate reconstructions from tree-rings available for Europe, Scandinavia and North Africa, there are very few reconstructions relating to the Middle Ages, one of the main reasons being the scarcity of continuous and reliable tree-ring series. The five longest temperature reconstructions covering the period 950–1500 are presented here. A sixth reconstruction is proposed which concerns the mean April to September temperature at the geographical point 45° N–10° E (Northeastern Italy), and a comparison is made with the five other reconstructions.

1. Introduction

Temperature reconstructions based on tree-ring widths or wood density are fairly numerous today in Europe and Scandinavia (Serre-Bachet, 1988; Serre-Bachet *et al.*, 1990); however, only a few of them extend back to the Medieval Epoch thus covering approximately the whole last millennium. The main reason for this is, of course, the scarceness of long continuous ring series obtained from still living, several hundred years old trees (Serre-Bachet, 1991). The long tree-ring series derived from archaeological material that are available for that period (Pilcher *et al.*, 1984) have often been established on the basis of miscellaneous wood samples of unknown origin, and hence they are unsuitable for reconstruction, with certain exceptions (Briffa *et al.*, 1990).

Almost all of the longest available reconstructions are based on tree-ring series relating to conifers (*Pinus, Picea, Larix*). However one reconstruction involves also deciduous tree-ring series (*Quercus*) and other proxy-data (documentary historical data and isotopic data) (Guiot *et al.*, 1988). These reconstructions concern mostly mean annual or monthly temperatures and they are mainly focused on the summer period.

Out of the main reconstructions available for Europe and Scandinavia, only six reach back to the Medieval Epoch, more or less covering it (Figure 1). The longest reconstruction shows the evolution of the April to August mean temperatures in Scandinavia from A.D. 500 to 1975 (Briffa *et al.*, 1990). Two reconstructions relate to the June to September mean temperature in Southeastern France and in Italy from A.D. 1150 to 1972 (Serre-Bachet and Guiot, 1987). The fourth

---

* Drs. Joel Guiot and Lucien Tessier kindly agreed to revise the manuscript after Dr. Serre-Bachet's death.

[†] Deceased 29 November 1992.

Climatic Change **26**: 213–224, 1994.
© 1994 *Kluwer Academic Publishers. Printed in the Netherlands.*

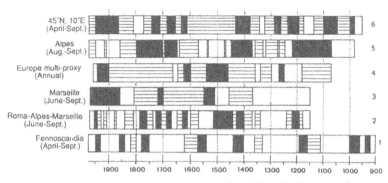

Fig. 1.   Climate reconstructions based on tree-rings back to 900 A.D. There are six reconstructions indicated by horizontal blocks. These reconstructions have a monthly to annual resolution. The coldest episodes are shown as bars with horizontal hatching and the warmest are in black. The length of each episode is either that defined by the authors or that inferred by us from the published curves. The white blocks correspond to climatic values close to the mean or with high frequency variations.

References and characteristics of these reconstructions
1. **Fennoscandia: Briffa** *et al.*, **1990.** April–August mean 'summer' temperature in Fennoscandia from A.D. 500 to 1975. Sixty-five standardized mean ring-width series and sixty-five maximum latewood densities series of living and remnant Scots pines (*Pinus silvestris* L.) as predictors, averages of Jones *et al.*'s (1985) gridded monthly temperature data as predictands.
2. **Rome-Marseille-Alps: Serre-Bachet and Guiot, 1987.** Mean summer temperature (June–September) common to Rome, Marseille and Grand St Bernard from A.D. 1972 back to 1150. Four mean ring-width series of larch (*Larix decidua* Mill.), fir (*Abies alba* Mill.) and pine (*Pinus leucodermis* Ant.) as predictors, temperature data of Rome, Marseille and Grand St Bernard pass as predictands.
3. **Marseille: Serre-Bachet and Guiot, 1987.** Main climatic periods for Marseille since A.D. 1150. Same data as 2.
4. **Western Europe: Guiot, 1988.** Annual mean temperature in Europe from A.D. 1070 to 1960, trend of 16 series at latitudes ranging from 40° N to 55° N. Eleven mean ring-width series of deciduous and coniferous trees plus nine other proxy series derived from historical archives (grape harvesting dates, thermal indices, temperature decadal estimates, wind frequency) and isotopic data (¹⁸O in the Arctic ice) as predictors, average of Jones *et al.*'s (1985) gridded monthly temperature data as predictands.
5. **Swiss Alps: Schweingruber** *et al.*, **1979, 1988.** Summer (August–September) temperature in the Swiss Prealps since A.D. 982 (982–1975) interpreted from latewood mean maximum densities of coniferous recent wood and construction wood from old buildings (Lauenen series). In the subalpine regions of the Alps, the high August and September temperatures influence the cell wall development in the latewood.
6. **44° N–10° E: Serre-Bachet, this paper.** April to September mean temperature from A.D. 970 to 1969 at the geographical point 45° N–10° E.

reconstruction is a synthesis of the annual mean temperatures of Europe from A.D. 1070 to 1960 (Guiot *et al.*, 1988); the fifth shows the evolution of mean August to September temperatures in the subalpine level of the Swiss Prealps from A.D. 982 to 1975 (Schweingruber *et al.*, 1979, 1988). Figure 1 summarizes the different markedly warm and cold episodes recorded in the five above-mentioned temperature reconstructions and in that presented in this paper. For the period A.D. 900–1500, there are a few instances of correspondences of short duration between

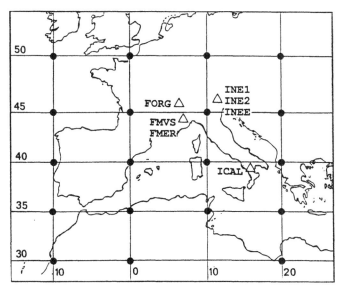

Fig. 2.   Geographical distribution of the seven tree-ring series used for April to September mean temperature reconstructions at the 19 Jones *et al.* (1985) grid-points for Southwestern Europe denoted by black dots on the map.

the different series; however, the 12th century, principally, appears as continuously warm on the Swiss series (5) and cool in series 1 and 4.

The present study is concerned with temperature reconstructions in Southwestern Europe and in Western Mediterranean on the basis of long tree-ring series established jointly by the Laboratoire de Botanique Historique et Palynologie of Marseille (France) and the Istituto Italiano di Dendrocronologia of Verona (Italy) (Serre-Bachet *et al.*, 1991).

2. Material and Method

Seven long tree-ring series were used, the geographical and main features of which are summarized in Figure 2 and Table I. Some series (FORG, FMVS, FMER and ICAL) were derived from cores sampled exclusively from living trees: larches (*Larix decidua* Mill), spruces (*Picea abies* L.) and pines (*Pinus leucodermis* Ant.); other series (INE1, INE2, INEE), principally in their oldest part, were derived from timber from buildings and artifacts. FMVS and FMER, as well as INE1 and INE2 may be regarded as partly redundant, because they were partly constructed from the same primary data. Use has been made of ring-width indices (Fritts, 1976) *i.e.* the ratio between the measured ring-width and the corresponding value of a 60-year low-pass filter applied to each data set.

A reconstruction could be made of the period 1362–1970 common to the seven series. Reconstructions were also possible for two other periods: 1150–1970 included in five of these series, and 970–1970 included in three series only. We

TABLE I: Characteristics of the seven tree-ring series used in reconstructions (**LADE**: *Larix decidua*, **PIAB**: *Picea abies*, **PILE**: *Pinus leucodermis*; precise or unknown (*m*) number of cores and pieces of wood averaged for the master chronology. Length of the reconstructed periods and corresponding number of tree-ring series (predictors) used in the reconstructions are shown at the bottom

FORG	(Orgère, 2300 m)	France	**LADE** 11	1353–1973
	(Tessier, 1986)			
FMVS	(Merveilles, 2150 m)	France	**LADE** 12	933–1974
FMER	(Merveilles, 2150 m)	France	**LADE** 8	100–1974
	(Serre, 1978)			
INE1	(Veneto, Trentino)	Italy	**LADE** m	781–1988
	(Alto-Adige, 2000 m)			
	(Bebber, 1990)			
INE2	(Same region as INE1)	Italy	**LADE** m	925–1984
	(Bebber, unpubl.)			
INEE	(Inst. It. Dendro.)	Italy	**PIAB** m	1362–1979
	(Martinelli, Pignatelli, unpubl.)			
ICAL	(Calabria, 1900 m)	Italy	**PILE** 18	1148–1974
	(Serre, 1985)			

Reconstructed periods:

7 series	1362–1969	(608 years)
5 series	1150–1969	(820 years)
3 series	970–1969	(1000 years)

chose to reconstruct the mean April to September temperature, in view of the geographical origin of the series, the altitude of the sites sampled, and the available set of temperature reconstructions, nearly all of them relating to the summer period. Temperature was thus reconstructed at 19 grid-points (Figure 2) according to Jones *et al.*'s (1985) gridded data. For each of these 19 grid-points, the transfer functions necessary for the intended reconstructions were calculated over the period 1851–1969, as the gridded data published by Jones *et al.* (1985) do not extend back farther than 1851. A lagged ring-width variable was included in the regression model; we therefore used a regression equation of the form

$$T_t = b_1 TRW_t + b_2 TRW_{t+1}$$

where T_t is the mean April–September temperature of a grid point in year t. TRW_t and TRW_{t+1} are the ring-width data of the years t and $t + 1$ respectively. Use was made of the *bootstrap* method (Efron, 1979) as adapted to the transfer function by Guiot (1989). In this method, which uses Monte Carlo simulation techniques, the original observations are sampled with replacement in a suitable way to construct pseudo-data sets (50 in the reconstruction presented here) on which the estimation of the statistical parameters are made. The method enables one to test the stability of the transfer function calculated both on the randomly sampled years, from which the calibration is made, and on the remaining years, used for verification. The reconstruction data retained for each year is the median of the 50 values available from the 50 pseudo-data sets. As with the Jones *et al.*'s data (1985), they are expressed as departure from the mean of the reference period 1951–1970, hence the term of 'anomalies' used in the following.

3. Results

Five of the 19 reconstructed points (Figure 2) provided reliable results suitable for reconstruction. Table II summarizes the results of the transfer functions obtained at the three points 45° N, 40° N, 30° N latitude and 10° E longitude. All the results are significant, but only point 45° N–10° E corresponding to Northeastern Italy will be discussed here. Indeed, at that point practically all the regressors used are involved in the reconstruction of the period 1362–1969, this being not true for the two other periods.

3.1. *The Characteristics of the Set of Reconstructed Temperature*

The curves (high frequency variations) of anomalies in the April to September temperatures at 45° N–10° E for the three periods studied are presented in Figure 3. The data estimated from the greatest number of regressors show the largest variation amplitude, which is quite normal since a greater part of the variance is thus explained; it also appears that for a good number of years, particularly 1370–1400 and 1670–1700, the same anomalies are recorded over the three sets of temperature reconstructions. High and significant correlation coefficients are obtained for the data that are common to each reconstructed series: **0.94** between the series starting in A.D. 970 and the series starting in 1150, **0.64** between those starting in 1150 and 1362, and **0.58** between those starting in 970 and 1362.

Examination of these same curves smoothed with a 25-year low-pass filter (low frequency variations) (Figure 4) shows more clearly the parallelism between the three curves. However, in the A.D. 1450–1500 interval, negative anomalies are more important in curve 1362–1969 than in the two others, and in the former the positive anomalies over A.D. 1740–1750 appear with a definite lag in comparison with the two other curves, where they are centered in 1720. Examination of indexed tree-ring series that have not been included in the temperature reconstruction over the two longest periods 1150–1969 and 970–1969, shows that these differences come from the ICA and INEE series (Table I). It is of interest to note the good parallelism between the filtered estimated and observed temperatures over the period 1851–1969. This underscores the quality of the reconstructions in the low-frequency domain.

3.2. *Comparisons*

Comparisons have been made between the April to September temperature reconstruction at grid-point 45° N–10° E (Figures 3 and 4) and other long reconstructions

Table II: Transfer function results at three grid points for each of the three periods available from the tree-ring series. Columns headed by m and m + 1 give the ratio of the regression coefficient to their standard deviation as calculated by the bootstrap procedure. Only significant values (0.05 level) are shown. A coefficient not significant at the 0.05 level is represented by its sign (+ or −). The last two columns give the multiple correlations calculated for the calibration and verification period. The bootstrap standard deviations are given in parentheses

		n							n + 1							Calibration	Verification
		INE1	INE2	FMVS	FMER	ICAL	FORG	INEE	INE1	INE2	FMVS	FMER	ICAL	FORG	INEE		
c	45N10E	1.98	+	−	+	−	+	3.89	−	+	3.01	3.82	2.08	2.25	−	0.598 (0.040)	0.372 (0.102)
	40N10E	2.26	2.88	−	−	−	2.31	+	−	−	2.23	2.46	2.28	+	−	0.559 (0.062)	0.366 (0.130)
	35N10E	2.21	3.24	−	−	+	2.24	+		−1.99	+	+	−	−	2.47	0.639 (0.038)	0.473 (0.082)
b	45N10E	+	+	+	+	−		−	−	+	3.94	3.98	+			0.473 (0.060)	0.310 (0.122)
	40N10E	+	2.37	−	−	+		+	+	−	2.40	2.59	+			0.471 (0.071)	0.315 (0.136)
	35N10E	2.08	2.13	+	+	+		−	−	−	2.22	2.42	−2.40			0.508 (0.042)	0.380 (0.087)
a	45N10E	2.44	+	+				+	+	+	3.63					0.412 (0.070)	0.329 (0.133)
	40N10E	2.32	2.81	−				−	+	−	2.26					0.438 (0.072)	0.353 (0.139)
	35N10E	2.98	3.21	+				+	+	−	+					0.444 (0.040)	0.375 (0.091)

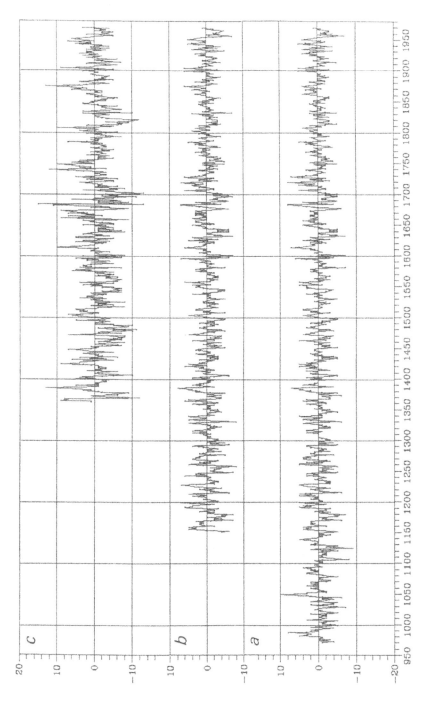

Fig. 3. Reconstructed 45° N–10° E temperature anomalies (1/10 °C) (April to September mean) relative to 1951–1970. Reconstruction on 3 (**a**: A.D. 970–1969) A.D.), 5 (**b**: A.D. 1150–1969) and 7 tree-ring series (**c**: A.D. 1362–1969).

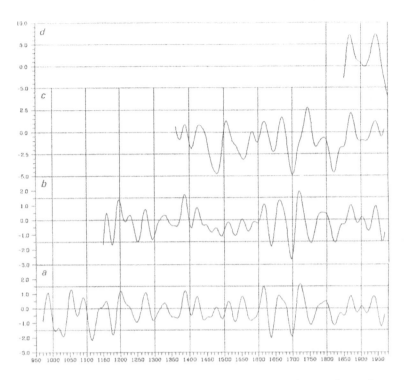

Fig. 4. Reconstructed April to September temperature anomalies (1/10 °C) at 45° N–10° E, smoothed with a 25-year low-pass filter (**a, b, c**, see Figure 3). **d:** Observed April to September temperature anomalies (Jones *et al.*, 1985) at the same point, filtered as before.

data available at the Marseille Laboratory. In comparison with the June to September temperature reconstruction at Rome since 1150 (Serre-Bachet and Guiot, 1987), the two unfiltered series starting at A.D. 970 and 1160, respectively, give correlation coefficients of **0.21** in comparison with the mean annual temperature reconstruction from 1070 at that same 45° N–10° E grid-point (Guiot *et al.*, 1988), based on a greater number of regressors (Figure 1, series 4), the series starting in A.D. 970 gives a coefficient of **0.17** and that starting in 1150 a coefficient of **0.09**. Nevertheless, all these low correlation coefficients are statistically significant ($P = 0.05$). As noted above, a period where the disagreement is obvious between Figure 4c, in one side, and Figure 4a–b, in the other side, is the 18th century: considering the tree-ring series involved in each reconstruction, it appears that the most northern site FORG is likely responsible for the discrepancy. Lack of correlation between larch growth in the northern (FORG) and southern (FMER) Alps have been noted previously (Guiot, 1982). A visual comparison with summer precipitation indices proposed by Alexandre (1987) for the interval A.D. 1000–1400 does not show any clear correspondences between the curves.

TABLE III: Main cold (negative anomalies) and warm (positive anomalies) episodes for the April to September temperature reconstruction (A.D. 970–1969) at 45° N–10° E

Very cold <-0.15 °C	Cold -0.15 to -0.05 °C	Warm 0.05 to 0.15 °C	Very warm >0.15 °C
		975–995	
995–1045			
		1045–1100	
1100–1135			
1170–1190			
		1190–1230	
	1230–1260		
		1260–1285	
	1285–1320		
	1340–1375		
		1375–1430	
	1430–1600		
			1600–1630
1630–1650			
		1650–1685	
1685–1710			
			1710–1740
	1740–1770		
	1805–1860		
		1860–1950	
	1950–1969		

3.3. *General trend*

Thorough examination of Figure 4a reveals that the amplitude of positive anomalies is on the whole smaller than the amplitude of the negative anomalies. Moreover, two types of mostly negative anomalies enable one to distinguish four major periods: (1) two periods, 970–1190 and 1600–1740, are characterized by anomalies near or exceeding 0.15 °C in absolute value and then by a high variability, (2) two periods, 1190–1600 and 1740–1969, are characterized by anomalies within about 0.10 °C, followed by low variability. If we consider Figure 4c more reliable since it is based on 7 tree-ring series, it appears that the high variability is a characteristic of the Little Ice Age from the late 1600s to the early 1800s. The period 970–1190 appears as a cold and highly variable period (mean anomaly = –0.32 °C) comparable to the Little Ice Age. The following period (1190–1450) is a warmer and more stable period. The most notable anomalies in these four periods are summarized in Table III and in Figure 1 (series 6). Only the Medieval Epoch (970–1500) will be discussed here.

4. Concluding Remarks about the Period A.D. 970–1500

It is clear from the temperature reconstruction at point 45° N–10° E (Figure 1, series 6) that the Medieval Warm Epoch, or Little Climatic Optimum, which is thought to have lasted at least from A.D. 1000 to 1300 (Lamb, 1984, 1988), was marked by as many cold as warm summer episodes. In particular no warm episode is recorded throughout the 12th century! Up to AD 1500, the climate was generally mild but only one really warm episode is recorded (at the end of the 14th century and the first three decades of the 15th century). The climate deterioration from A.D. 1300 reported by several authors (Lamb, 1984, 1988; Le Roy Ladurie, 1967; Steenberg, 1951) for northern Europe is not really found for the location of gridpoint 45° N–10° E.

Detailed comparison between the reconstruction obtained at gridpoint 45° N–10° E and the five other temperature reconstructions available for the period A.D. 970–1500 (Figure 1) reveals some correspondences in the course of the last 10 centuries. The following periods are common to at least two of the seven series compared: end of the **10th century:** *warm* (45° N–10° E and Fennoscandia); last three decades of the **11th century:** *warm* (45° N–10° E and Swiss Alps); first three decades of the **12th century** (45° N–10° E, Fennoscandia and western Europe) and end of that century (45° N–10° E and Rome-Marseille-Alps): *cold*; beginning of the **13th century:** *warm* (45° N–10° E, Rome-Marseille-Alps and Swiss Alps) and second half of that century: rather *cold*, with a warm episode in the last three decades (45° N–10° E, western Europe and Swiss Alps); beginning of the **14th century** (45° N–10° E, western Europe and Swiss Alps) and middle of the century (45° N–10° E, Fennoscandia, Rome-Marseille-Alps and Swiss Alps): *cold*; **end of the 14th century:** *warm* (45° N–10° E and Swiss Alps); first three decades of the **15th century:** *warm* (45° N–10° E, Fennoscandia, Rome-Marseille-Alps and Swiss Alps); middle of the century (45° N–10° E, Marseille and western Europe) or beginning of the second half of that century (45° N–10° E and Swiss Alps): *cold*.

These episodes, that are common to the temperature series reconstructed from totally independent tree-ring sequences (45° N–10° E, Fennoscanda and Swiss Alps) derived from different European areas, are not fortuitous; they reflect broader climatic trends that prevailed in western Europe and Scandinavia during the first half of the last millennium. The relation between some of these episodes and the series corresponding to the mean annual temperature variations throughout Europe (Figure 1, series 4) is probably related to the tree-ring series used in that reconstruction, these series mostly depending on the temperatures of the tree-growth period which covers approximately the April to September period.

Acknowledgements

I am very grateful to Anna Bebber, Nicoletta Martinelli and Olivia Pignatelli, from the Istituto di Dendrocronologia di Verona in Italy, for supplying their longest Veneto and Trentino tree-ring series mainly built under E.E.C. grant 3528/86 and CEE/CNR contract EV4C-0082-I.

References

Alexandre, P.: 1987, *Le Climate en Europe au Moyen Age*, Ed. Ecole des Hautes Etudes en Sciences Sociales Paris, 827 pp.

Bebber, A. E.: 1990, 'Una Cronologia di Riferimento del Larice (*Larix decidua* Mill.) delle Alpi Orientali Italiane', *Dendrochronologia* **8**, 119–139.

Briffa, K. R., Bartholin, T. S., Eckstein, D., Jones, P. D., Karlen, W., Schweingruber, F. H., and Zetterberg, P.: 1990, 'A 1,400-Year Tree-Ring Record of Summer Temperatures in Fennoscandia', *Nature* **346**, 434–439.

Efron, B.: 1979, 'Bootstrap Methods: Another Look at the Jackknife', *Ann. statis.* **7**, 1–26.

Fritts, H. C.: 1976, *Tree-Rings and Climate*, Academic press, New York, 567 pp.

Guiot, J.: 1982, 'Deux Méthodes de l'Utilisation de l'Épaisseur des Cernes Ligneux pour la Reconstruction de Paramètres Climatiques Anciens; l'Exemple de Leur Application dans le Domaine Alpin', *Paleogeogr., Paleclim., Paleoecol.* **45**, 347–368.

Guiot, J.: 1989, 'Methods of Calibration', in Cook, E. R. and Kairiukstis, L. A. (eds.), *Methods of Dendrochronology: Applications in the Environmental Sciences*, Kluwer Academic Publishers, pp. 165–178.

Guiot, J., Tessier, L., Serre-Bachet, F., Guibal, F., Gadbin, C.: 1988, 'Annual Temperature Changes Reconstructed in W-Europe and NW-Africa back to A.d. 1100', *Annal. Geophys.* special issue, XIII General Assembly of European Geophysical Society, Bologna, 21–25 March 1988, p. 85.

Jones, P. D., Raper, S. C. B., Santer, B. D., Cherry, B. S. G., Goodess, C., Bradley, R. S., Diaz, H. F., Kelly, P. M., and Wigley, T. M. L.: 1985, 'A Grid Point Surface Air Temperature Data Set for the Northern Hemisphere, 1851–1984', DOE Tech. Rep. TRO22 (U.S. Department of Energy, Washington DC), 251 pp.

Lamb, H. H.: 1984, 'Climate in the Last Thousand Years: Natural Climatic Fluctuations and Change', in Flohn, H. and Fantechi, R. (eds.), *The Climate of Europe: Past, Present and Future*, Reidel Publishing Company, pp. 25–64.

Lamb, H. H.: 1988, *Weather, Climate and Human Affairs*, Routledge London, 364 pp.

Le Roy Ladurie, E.: 1967, *Histoire du Climat depuis l'An Mil*, Flammarion Paris, 366 pp.

Pilcher, J. R., Baillie, M. G. L., Schmidt, B., and Becker, B.: 1984, 'A 7,272-Year Tree-Ring Chronology for Western Europe', *Nature* **312**, 150–152.

Richter, K. and Eckstein, D.: 1990, 'A Proxy Summer Rainfall Record for Southeast Spain Derived from Living and Historic Pine Trees', *Dendrochronologia* **8**, 67.

Schweingruber, F. H., Bräker, O. U., and Schär, E.: 1979, 'Dendroclimatic Studies on Conifers from Central Europe and Great Britain', *Boreas* **8**, 427–452.

Schweingruber, F. H., Bartholin, T., Schär, E., and Briffa, K. R.: 1988, 'Radiodensitometric-Dendroclimatological Conifer Chronologies from Lapland (Scandinavia) and the Alps (Switzerland)', *Boreas* **17**, 559–566.

Serre, F.: 1978, 'The Dendroclimatological Value of the European Larch (*Larix decidua* Mill.) in the French Maritime Alps', *Tree-Ring Bull.* **38**, 25–34.

Serre-Bachet, F.: 1985, 'Une Chronologie Pluriséculaire du Sud de l'Italie', *Dendrochronologia* **3**, 45–66.

Serre-Bachet, F.: 1988, 'La Reconstruction Climatique à Partir de la Dendroclimatologie', *Public. Assoc. Internat. Climatol.* **1**, 225–233.

Serre-Bachet, F.: 1991, 'Tree-Rings in the Mediterranean Area. Evaluation of Climate Proxy Data in Relation to the European Holocene', in Frenzel, B. (ed.), *Paleoclimate Research*, pp. 133–147.

Serre-Bachet, F. and Guiot, J.: 1987, 'Summer Temperature Changes from Tree-Rings in the Mediter-
    ranean Area during the Last 800 Years', in Berger, W. H. and Labeyrie, L. D. (eds.), *Abrupt
    Climatic Change. Evidence and Implications*, Reidel Pub. Co., pp. 89–97.
Serre-Bachet, F., Guiot, J., and Tessier, L.: 1990, 'Dendroclimatic Evidence from Southwestern
    Europe and Northwestern Africa', in Bradley, R. S. and Jones, P. D. (eds.), *Climate since A.D.
    1500*, Routledge London, pp. 349–365.
Serre-Bachet, F., Martinelli, N., Pignatelli, O., Guiot, J., and Tessier, L.: 1991, 'Evolution des
    Températures du Nord-Est de l'Italie depuis 1500 A.D. Reconstruction d'après les Cernes des
    Arbres', *Dendrochronologia* **9**, 213–229.
Steenberg, A.: 1951, 'Archaeological Dating of the Climatic Change in North Europe about A.D.
    1300', *Nature* **168**, 672–675.
Tessier, L.: 1986, 'Chronologie de Mélèzes des Alpes et Petit Age Glaciaire', *Dendrochronologia* **3**,
    97–113.
Till, C. and Guiot, J.: 1990, 'Reconstruction of Precipitation in Morocco since A.D. 1100 Based on
    Cedrus atlantica Tree-Ring Widths', *Quatern. Res.* **33**, 337–351.

(Received 1 June 1993; in revised form 15 November, 1993)

THE MEDIEVAL WARM PERIOD ON THE SOUTHERN COLORADO PLATEAU

JEFFREY S. DEAN

Laboratory of Tree-Ring Research, Building 58, The University of Arizona, Tucson, Arizona 85721, U.S.A.

Abstract. Several questions concerning the Medieval Warm Period (MWP), an interval (A.D. 900 to 1300) of elevated temperatures first identified in northern Europe, are addressed with paleoenvironmental and archaeological data from the southern Colorado Plateau in the southwestern United States. Low and high frequency variations in alluvial groundwater levels, floodplain aggradation and degradation, effective moisture, dendroclimate, and human adaptive behavior fail to exhibit consistent patterns that can be attributed to either global or regional expressions of the MWP. There is some suggestion, however, that climatic factors related to the MWP may have modified the regional patterns to produce minor anomalies in variables such as the number of intense droughts, the occurrence of specific droughts in the twelfth and thirteenth centuries, the prevalence of low temporal variability in dendroclimate, and the coherence of some low and high frequency environmental variables and aspects of human adaptive behavior. These results suggest that the MWP does not represent warming throughout the world. Rather, it was a complex phenomenon that probably was expressed differently in different regions.

1. Introduction

The Medieval Warm Period (MWP), or Little Climatic Optimum, is viewed as an interval of above average temperatures that lasted from approximately A.D. 900 to 1300 (Lamb, 1977). This climatic excursion is defined primarily on the basis of historical observations and proxy climate records in the British Isles and northern Europe. Several important questions relating to this phenomenon were addressed at the Workshop on the Medieval Warm Period in Tucson, Arizona, in 1991 (Diaz and Hughes, 1991). Did an MWP actually exist and, if so, what were its geographic and chronological parameters? Was it a global (or at least hemispherical) phenomenon, or was it restricted to Europe? Did it span the entire period defined in Europe, or was it shorter or longer in other areas? Was it continuous or discontinuous? Was it everywhere characterized by elevated temperatures, or was it expressed differently in different areas?

One approach to these questions is to inspect a broad range of historical and proxy climate data for evidence of anomalies that might either disclose a uniform global signal or identify local expressions of the larger phenomenon. The absence of such anomalies would, of course, cast doubt on the 'universality' of this climatic episode. Another approach is to examine archaeological data for human behavioral transformations that might be related to long-term environmental changes like the MWP. Due to an abundance of high quality paleoenvironmental and archaeological

Climatic Change **26**: 225–241, 1994.

information, the southern Colorado Plateau in the United States Southwest should provide a good, regionally specific test of the MWP hypothesis. The purpose of this paper is to examine extant Plateau paleoenvironmental and human behavioral records for properties that might be related, at least conceptually, to the climatic anomalies that characterize the MWP in Europe.

2. Theoretical and Methodological Considerations

The preferred approach to this task would be to use known climatic relationships in the Northern Hemisphere to develop deductive expectations about Colorado Plateau climate during the 10th through 13th centuries from the characteristics of the MWP in Europe. The hypothesis of a hemisphere-wide MWP could then be strengthened or weakened through a search of the Plateau paleoenvironmental records for indicators of the expected anomalies. Unfortunately, the global climate system is as yet too poorly understood to allow aspects of Plateau climate to be predicted from conditions in the North Atlantic region. Therefore, a different, somewhat less ideal tack must be taken. This tactic involves, first, an examination of Plateau paleoenvironmental data for fluctuations that might plausibly be equated with the anomalies evident in Europe. Second, the record of human adaptive behavior in the region is inspected for configurations and changes that might have been stimulated by environmental factors related to the MWP. Equating Plateau paleoenvironmental variability and past human behavior with the MWP rests on correspondences in direction (or explainable differences in direction) and/or similarities in time span and relative intensity of these variations.

Although the MWP is recognized in northern Europe as a climatic phenomenon, there are sound reasons for not limiting the Colorado Plateau test solely to past climatic variability. Many nonclimatic environmental processes and indicators are influenced by climate. Geologic and hydrologic variables such as streamflow, the rise and fall of alluvial water tables, and the deposition and erosion of floodplain sediments may be related to climate and may therefore be used as indicators of past climatic fluctuations. Plant pollen production and changes in the elevational distributions of plant communities are related to available moisture (Hevly, 1988) and often reflect changes in climate. Human behavior, particularly in the arid Southwest, is constrained by climatic conditions and changes and by fluctuations in other, climate-related environmental factors (Baerreis *et al.*, 1976; Butzer, 1982). Therefore, evaluation of sensitive aspects of past human behavior (particularly subsistence, settlement, and interaction) should shed light on the environment of the target period.

Seventy years of research has produced an extensive and varied Colorado Plateau paleoenvironmental record (Gumerman, 1988). Based on the natural processes that regulate environmental fluctuations and the differential sensitivities of the indicators of these variables, two classes of environmental variability are recognized, those due to low frequency natural processes and those due to high frequency processes (Dean, 1988a, b). Scaled to human behavioral dimensions by reference to one human generation, low frequency variability is that due to processes with periodicities, either regular or irregular, greater than or equal to 25 yr, while high frequency variability is that due to processes with cycles of less than 25 yr.

Three attributes of the low frequency-high frequency dichotomy should be emphasized. First, the typology refers to the processes that govern environmental variability not to the outcomes, which can change at rates incommensurate with the periodicities of the controlling processes. For example, low frequency natural processes can produce abrupt environmental changes, as when arroyos propagate rapidly through drainage systems. Second, the classes represent end points on a continuum of potential variability; most natural processes oscillate across a broad spectrum of frequencies as do the environmental outcomes of their operation. Third, the dichotomy is as much a function of the temporal sensitivity of various paleoenvironmental reconstruction techniques as it is of the intrinsic properties of the processes.

Environmental fluctuations caused by low frequency natural processes are indicated by several different types of evidence. Chronostratigraphic studies of floodplain deposits specify the rise and fall of alluvial groundwater levels, the deposition and erosion of alluvium, the stabilization and disturbance of alluvial surfaces, and vegetation changes on these surfaces. Palynology, the analysis of macrobotanical samples from archaeological sites, and packrat midden studies reveal changes in the elevational and habitat distributions of plant species and communities. Some dendroclimatic analyses provide information on low frequency climatic trends. There are fewer indicators of high frequency environmental variations. High resolution palynological analyses disclose rapid fluctuations in pollen production by various species, but such studies are relatively rare. The most direct source of data on high frequency paleoclimatic variability is the dendroclimatic analysis of tree growth.

3. Southern Colorado Plateau Paleoenvironmental Records

With the foregoing potentials and limitations in mind, Colorado Plateau paleoenvironmental records are surveyed for possible indicators of MWP anomalies. Due to the disparate sensitivity of the various paleoenvironmental techniques, this appraisal of the region's environmental history during the last millennium is organized in terms of low and high frequency process variables.

3.1. *Low Frequency Process Environmental Variability*

Because fluctuations in fluvial conditions are thought to be caused by climatic factors (Karlstrom, 1988), the record of low frequency fluvial variations can be examined for attributes that may be related to the Medieval Warm Period. Past hydrologic and depositional variability on the Colorado Plateau is indicated by

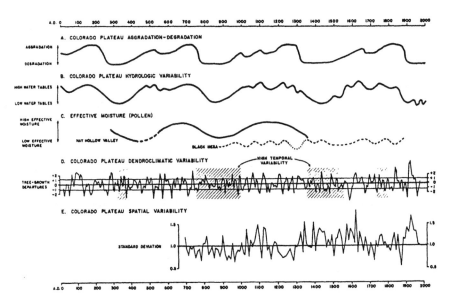

Fig. 1. Low and high frequency process environmental variability on the southern Colorado Plateau during the last two millennia.

a series of chronostratigraphic studies of alluvium in northeastern Arizona by Gregory (1916, 1917), Antevs (Sears, 1937), Hack (1942, 1945), Thornthwaite *et al.* (1942), Cooley (1962), and the Karlstroms (Karlstrom, 1983; Karlstrom and Karlstrom, 1986; Karlstrom, 1988; Karlstrom *et al.*, 1976). These analyses reveal a fairly regular cycle of alluvial hydrologic fluctuations (Figure 1B; Dean *et al.*, 1985, Figure 1; Euler *et al.*, 1979, Figure 4; Karlstrom, 1988; Plog *et al.*, 1988, Figure 1) with a primary periodicity of approximately 550 yr and a spectrum of shorter oscillations (Karlstrom, 1988, p. 64). Due to threshold relationships between groundwater variability and sedimentation, the depositional sequence has an asymmetrical profile (Figure 1A). The MWP (A.D. 900–1300) overlaps with a period of rising water tables that lasted from approximately 850 to the middle 1100s and an interval of general alluviation that extended from about 925 to around 1275 when arroyo cutting began.

Geologic-hydrologic records from elsewhere on the Colorado Plateau lack the time resolution of the northeastern Arizona sequences. Nevertheless, the former exhibit general trends similar to those of the latter (Euler *et al.*, 1979, Figure 4; Karlstrom, 1988, pp. 65–70). Eddy (1964) argues that the interval between 800 and 1550 along the San Juan River in northwestern New Mexico was characterized by nondeposition, channel entrenchment, and terrace stability. Poor chronological control due to the lack of associated archaeological sites postdating 900, however, allows the possibility that deposition continued through the MWP here as it did farther west. Deposition prevailed in Chaco Canyon until after the end of the Bonito phase (Bryan, 1954; Hall, 1977, 1983; Love, 1980) and thus could have persisted into the 13th century (Love, 1980, Figure 85). In the Grand Gulch area of southeastern Utah (Lipe and Matson, 1975), deposition continued through Pueblo III times; that is, until after 1250.

To summarize, the period from 900 and 1300 was characterized primarily by alluviation along the drainages of the Colorado Plateau. It is tempting to relate this depositional episode to some sort of regional climatic divergence linked through the global climate system to the conditions that distinguish the MWP in Europe. Unfortunately, this inference is contradicted by the fact that sedimentation regularly alternated with erosion during the last four millennia. Indeed, hydrologic variability on the Colorado Plateau during the last 2000 yr (Figure 1B) exhibits an internal dynamic that is unlikely to be related to the MWP or any single climatic excursion. It seems, therefore, that low frequency fluvial variability between 900 and 1300 cannot be ascribed to a Southwestern expression of a global climatic anomaly that produced the MWP in Europe.

While fluctuations in fluvial systems cannot be related to the MWP, low frequency variations in vegetation as revealed by numerous pollen studies (Bohrer, 1972; Buge and Schoenwetter, 1977; Fall *et al.*, 1981; Hall, 1977, 1983; Hevly, 1964; Schoenwetter, 1962, 1966, 1967, 1970) might reflect this climatic excursion. Many of these palynological reconstructions are less useful in this regard than might be expected. A chief drawback has been the use of samples from archaeological contexts to maximize preservation and chronological control (Schoenwetter, 1962). Unfortunately, the environmental signal of such samples is distorted by human activity at the archaeological sites (Hevly, 1981, 1988; Kelso, 1976). Samples from natural contexts generally provide less biased environmental information but at the cost of lower temporal resolution. In many cases, combinations of the two sources must be used, with expectably mixed results.

Inspection of pollen records from the Plateau (Hevly, 1988) reveals no consistent local changes in either vegetation or effective moisture over the entire 900–1300 period. Many areas, however, exhibit high effective moisture between 1000 and 1200, roughly the middle of the MWP. Petersen's (1988) integrated study of several paleoenvironmental indicators, including pollen, is the most comprehensive of these analyses. Elsewhere in this issue he summarizes his evidence for warmer and wetter conditions during the middle of the MWP in southwestern Colorado. Whether these pollen spectra reflect the global climatic processes that produced the MWP in Europe is unknown. The marked similarity between the hydrologic curve (Figure 1B) and low frequency variation in effective moisture (Figure 1C) suggests that both these indicators reflect regional environmental processes that vary independently of climatic deviations like the MWP.

Additional information on low frequency environmental variability comes from the analysis of pollen and macrobotanical remains in packrat middens (Betancourt, 1990). Comparatively low tempral resolution, however, makes it difficult to relate midden based reconstructions to relatively brief and recent climate events such as the MWP. Some vegetation changes that occur during the MWP, such as the disappearance of pinyon from the Chaco Canyon record after 900 (Betancourt *et al.*, 1983, Figures 3 and 4; Betancourt and Van Devender, 1981, Figure 1), appear to be anthropogenic rather than climatic in origin (Betancourt *et al.*, 1983, 211–213). At present, Southwestern packrat midden analysis seems most sensitive to broad scale vegetation changes and has little relevance to short, low amplitude deviations like the MWP.

Efforts to extract low frequency components from Plateau tree-ring chronologies have generally foundered on the low sensitivity of tree growth to long term climatic variations and the analytical removal of trends from the ring series (Dean, 1988b, pp. 151–154). Nevertheless, the tree-ring record does possess some features that are potentially relevant in this context. Because dendrochronology pertains primarily to rapid fluctuations, however, these aspects are considered below with the rest of the tree-ring data.

Given the disparate measures of low frequency environmental variability, the various reconstructions exhibit a remarkable coherence during the 900–1300 period. Collectively, they show rising alluvial groundwater levels, floodplain accretion, increased effective moisture, and a widening of the farmable belt in southwestern Colorado during the 900 to 1150 subinterval. These factors peaked in the middle 12th century and declined thereafter, a delay in the onset of arroyo cutting being due to systematic lag effects. Viewed in the context of long-term environmental variability in the region, however, these correspondences cannot unequivocally be attributed to climatic anomalies such as the MWP. Most of these factors varied synchronously both before and after the MWP to define a long term cyclic pattern. Given the persistence of this pattern, it seems unlikely that low frequency variability between 900 and 1300 can be attributed to a global climatic phenomenon limited to that interval. On the other hand, the Plateau pattern raises the possibility that the MWP is a European manifestation of global processes that produced the long-term Southwestern cycle of environmental change.

3.2. *High Frequency Process Environmental Variability*

A few high resolution pollen sequences (Fall *et al.*, 1981; Hevly, 1988) provide information on rapid environmental variability on the Colorado Plateau. Since these fluctuations generally parallel those of the dendroclimatic record (Figure 1C; Dean *et al.*, 1985, Figure 1; Euler *et al.*, 1979, Figure 4; Plog *et al.*, 1988, Figure 8.1), they are not considered further here.

Tree-ring research on the southern Colorado Plateau has produced qualitative reconstructions of relative dendroclimatic variability at 25 climate sensitive tree-growth stations (Figure 2; Dean and Robinson, 1978) and quantitative retrodictions of variables such as seasonal and annual precipitation, seasonal temperatures, and Palmer Drought Severity Indices (PDSI) for the middle Rio Grande valley (Rose *et al.*, 1981), the Four Corners area (Rose *et al.*, 1982; Van West, 1990), and Black Mesa (Dean, 1986; Lebo, 1991). In addition, tree-ring analysis has been used to

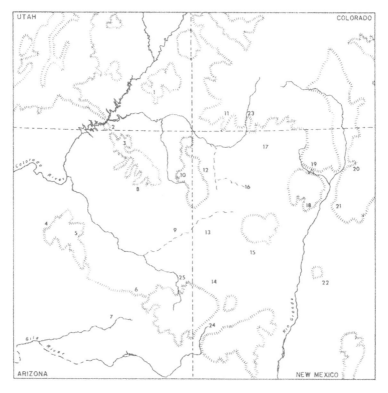

Fig. 2. Expanded, climate sensitive tree-ring chronologies on the southern Colorado Plateau: 1, Natural Bridges; 2, Navajo Mountain; 3, Tsegi Canyon; 4, Coconino Plateau; 5, Flagstaff; 6, Central Mountains North; 7, Central Mountains South; 8, Hopi Mesas; 9, Puerco Valley; 10, Canyon de Chelly; 11, Mesa Verde; 12, Chuska Valley; 13, Cibola; 14, Quemado; 15, Cebolleta Mesa; 16, Chaco Canyon; 17, Gobernador; 18, Jemez Mountains; 19, Chama Valley; 20, Rio Grande North; 21, Santa Fe; 22, Chupadero Mesa; 23, Durango; 24, Reserve; 25, Little Colorado.

reconstruct annual crop yields in southwestern Colorado (Burns, 1983; Van West, 1990), Black Mesa (Lebo, 1991), and the Pajarito Plateau (Garza, 1978).

Inspection of the 25 qualitative reconstructions (Dean and Robinson, 1977) reveals that relative dendroclimatic variability during the 900–1300 period did not diverge systematically from that of other intervals. Thus, there appear to be no long term deviations from average tree growth at any of these localities that can be linked to the MWP.

In general, the quantitative reconstructions for the middle Rio Grande Valley (Rose *et al.*, 1981) and the Four Corners (Rose *et al.*, 1982) also lack long term deviations that span the entire MWP. A possible exception to this observation is found in the June PDSI reconstruction for northwestern New Mexico (Dean, 1992; Rose *et al.*, 1982), which reveals a severe summer drought between 1130 and 1180 that coincides with the transition from ameliorating to worsening hydrologic and effective moisture conditions. This co-occurrence of low and high frequency environmental changes may be a local manifestation of global climatic processes related to the Medieval Warm Period in Europe.

Burns' (1983) and Van West's (1990) crop yield reconstructions for southwestern Colorado indicate reduced production during much of the 12th century, a span that is coeval with the summer drought in northwestern New Mexico and the transition from favorable to unfavorable low frequency environmental conditions. Farther west, a reconstruction of annual precipitation on Black Mesa (Dean, 1986) shows a lesser negative excursion in the mid-1100s than do the more easterly retrodictions. One feature common to all the qualitative and quantitative reconstructions is an intense dry interval at the end of the MWP, the 'Great Drought' of 1276–1299 (Douglass, 1929, 1935; Haury, 1935).

Although the individual dendroclimatic records fail to exhibit consistent patterns over the entire 900–1300 interval, integrating dendroclimatic variability across the Colorado Plateau may reveal larger scale aspects of high frequency variability that may be equated with the MWP. Figure 1D illustrates the decade departures for a composite tree-ring chronology created by averaging 21 local climate-sensitive ring series. The values are calculated for successive decades overlapped by five years and are expressed in standard deviation units that represent positive and negative departures from the mean of the composite chronology. The departures during the 900–1300 interval do not differ appreciably in amplitude variability from those of other periods. The MWP is, however, bracketed by a longer interval (roughly 750 to 1550) characterized by a relative dearth of intense and prolonged droughts.

Not surprisingly, the composite Colorado Plateau amplitude series resembles the locality sequences in lacking a strong pattern for the entire span of the MWP. Shorter intervals within this period, however, do exhibit some fairly unusual deviations from long term means that may be Southwestern manifestations of global climatic processes linked to the MWP. These include the comparative paucity of intense droughts, the major drought of the mid-1100s, and the Great Drought at the end of the 13th century. Unfortunately, we as yet have no general model of northern hemisphere climatic systematics that allows the Southwestern dendroclimatic amplitude phenomena to be processually related to the European MWP.

Other aspects of Plateau high frequency dendroclimatic variability can be examined for concordance with the Medieval Warm Period. The oscillations of the regional departures exhibit two contrasting temporal patterns (Figure 1D). Part of the record exhibits relatively gradual transitions from maximum to minimum values. At other times, the changes from high to low amplitudes were much more rapid, usually occurring from one decade to another. These intervals of high temporal variability are indicated by hatching on Figure 1D. Except for the 10th century, the MWP is characterized by low temporal variability with fairly broad swings from high to low departures. That this behavior is not due to global climatic conditions unique to the MWP is shown by the prevalence of low temporal variability during four other intervals (Figure 1D). Nonetheless, the possibility that low temporal variability reflects global climate conditions characteristic of the MWP deserves further investigation.

Spatial variability in dendroclimate (Figure 1E) is measured by the standard deviation of the station departures for each decade. Low values specify comparative uniformity in climate across space, while high values denote major differences across space. The MWP exhibited the full range of spatial variability, which was low at the beginning, increased to a peak in the middle 12th century, and declined to a minimum at the end. Thus, this variable exhibits no particular correlation with the MWP, although the interval of high values centered on 1100 may be related to the other low and high frequency factors that covary during the middle of the MWP.

As was the case with low frequency environmental factors, high frequency variables, either singly or collectively, exhibit no consistent patterns unique to the entire 400-yr span of the MWP. On the other hand, a few regularities may reflect broad scale climatic processes that relate Southwestern climate to that of Europe during the Medieval Warm Period. Among these factors are the rarity of intense droughts, the predominance of low temporal variability, a prolonged drought in the mid-12th century, and the Great Drought at the end of the 13th century. These ambiguous results indicate that the Southwestern consequences of the global climatic processes that produced the MWP were complex and are not easily discriminated by the various paleoenvironmental techniques.

4. Human Behavior on the Southern Colorado Plateau, A.D. 900–1300

Located near the northern limit of reliable maize cultivation in western North America, the Colorado Plateau is marginal for agricultural subsistence. Therefore, Anasazi adaptive behavior must have been attuned to both the inhospitable general environment and to any fluctuations in the conditions that limited agricultural production (Dean, 1988a). Because of this sensitivity, aspects of Anasazi behavior should reflect environmental variations related to the Medieval Warm Period. Many summaries of the prehistory of the Colorado Plateau are available (Cordell, 1984; Cordell and Gumerman, 1989), and it would be pointless to provide a synthesis here. Instead, relevant adaptive aspects of the human occupation of the Colorado Plateau during the MWP are summarized. Three particularly sensitive aspect of human adaptive behavior –subsistence, settlement, and interaction – are emphasized.

Subsistence varied across space and time during the MWP. Anasazi populations had access to a limited but versatile agricultural technology whose elements were differentially activated in response to particular adaptive situations. Thus, the Mesa Verdeans relied on dry farming in a productive upland zone (Petersen, this issue), the Chacoans combined sophisticated water control systems with a large scale trade network in a less productive lowland setting (Vivian, 1990), and the Kayentans applied a variety of small scale farming techniques to a varied but agriculturally limited environment. Constant adjustments were necessary to attune the mix of techniques to fluctuating conditions, and no uniform regional or local subsistence responses to a general MWP environmental anomaly can be identified.

Settlement – which is reflected in the distribution of sites across the landscape, their internal organization, and their relationships – is a diverse and flexible adaptive mechanism that is sensitive to environmental variation. Interaction within and among communities is a primary adaptive strategy that is closely related to settlement. Obviously, the multifarious ramifications of these complex issues cannot be addressed in detail here. Rather, the broad picture of Anasazi settlement and interaction across the Plateau between 900 and 1300 may indicate some general responses to environmental variability during the MWP.

The first part of the MWP was a time of major growth in population, geographical range, organizational complexity, and intergroup interaction on the Plateau. During the 900–1150 period, the Anasazi occupied every habitable locale and expanded outward to achieve their maximum range, which extended from central Utah south to the Mogollon Rim and from the Virgin River east to the Rio Grande. At the same time, a regional interaction system centered on Chaco Canyon expanded beyond the San Juan Basin and achieved an unprecedented level of sociocultural integration and complexity. After 1150, however, Anasazi culture underwent major changes. The total range began to shrink as people either abandoned the peripheral areas or adopted lifeways no longer recognizable as Anasazi. Populations concentrated in localities particularly favorable for farming. The core of the Chacoan regional system 'collapsed' leaving successor systems in the Mesa Verde and Cibola areas. Major reorganizations and realignments throughout the region produced social systems quite different from those that existed previously. Specific settlement configurations varied across space, but most areas were characterized by large communities and interaction networks of variable extent. These developments ended by 1300 when the San Juan drainage was abandoned by the Anasazi who moved south and east to join related populations in areas now occupied by Puebloan peoples.

The causes of Anasazi cultural developments during the MWP are undoubtedly complex and involve numerous interconnected social, cultural, environmental, and historical factors. It probably is no accident, however, that the advancements of the 900–1150 period coincided with the most propitious environmental epoch of the last 2,000 yr (Plog *et al.*, 1988). Alluvial water tables were rising, floodplains were aggrading, effective moisture was increasing, severe droughts were rare, and spatial variability in climate (which facilitated interareal interaction) was increasing. Early in the period, amplitude variation in precipitation was low, and temporal variability was high, both of which favored the accumulation of stored food reserves to offset production shortfalls (Jorde, 1977). After 1000, amplitude variation increased, and high temporal variability gave way to longer trends, conditions that enhanced the predictability of crop yields. This combination of environmental factors facilitated the occupation of a wide range of habitats, population growth, large scale range expansion, interareal exchange, and sociocultural complexity that characterized the 900–1150 period.

The mid-12th century population redistributions and social reconfigurations occurred during a period of stress caused by the interaction of demographic, socio-cultural, and environmental factors. High population levels strained the capacities of local groups to sustain themselves, and the delicately balanced Anasazi organizational systems were especially vulnerable to perturbations in any components of their adaptive situations. At this critical juncture, these societies were faced with serious environmental problems (Figure 1). A secondary hydrologic decline lowered alluvial water tables and made floodplain farming more dependent on rainfall, a far less reliable source of moisture. At the same time, a severe drought decreased the amount of atmospheric water available for farming. In addition, declining spatial variability in climate reduced the production differentials that made exchange a viable way of offsetting local subsistence deficits. The combination of critical population densities, low and high frequency environmental deterioration, and limited sociotechnological response capabilities undoubtedly was a major factor in the large scale settlement and organizational changes that occurred across the Anasazi domain in the latter half of the 12th century.

The reorganizations of Anasazi societies after 1150 occurred during a brief interlude of environmental conditions that, though not unfavorable, were different from those of the 900–1150 period. During the early 1200s, high groundwater levels and alluviation, fairly high annual rainfall, and low temporal variability in climate facilitated agriculture. On the other hand, low spatial climatic variability inhibited the reestablishment of regional networks like the Chacoan system and fostered interareal differences. The increased spatial variability in culture and the absence of a single dominating regional interaction system undoubtedly reflect adaptations to altered environmental, demographic, and societal circumstances that favored local food production and constrained the exchange of subsistence goods.

The Anasazi abandonment of the San Juan drainage during the last quarter of the 13th century coincided with major low and high frequency environmental transformations. Falling alluvial groundwater levels and rapid arroyo cutting associated with a first order hydrologic minimum required greater dependence on precipitation for agricultural production. At the same time, the Great Drought reduced the amount of water available from rainfall. The persistence of low temporal climatic variability bolstered expectations of further deterioration in the subsistence system. Continuing spatial uniformity limited exchange as a means of compensating for local food shortages. Demographic and organizational constraints prevented local behavioral adjustments to the worsening adaptive situation (Dean, 1966, 1969, 1987), and the Anasazi vacated the San Juan drainage as a locus of permanent residence and moved to areas where conditions allowed the survival of their social systems. This relocation coincided with the end of the MWP, and Petersen (this issue) argues that the Anasazi failed to reoccupy the San Juan drainage because ensuing Little Ice Age conditions inhibited farming in the Four Corners area. It should be stressed, however, that socioreligious developments farther south exerted a powerful attractive force that reinforced the environmental 'push' away from the San Juan drainage (Adams, 1991; Dean, 1991).

The chronicle of human adaptive behavior on the Colorado Plateau between 900 and 1300 reveals no uniform pattern that can be attributed to a single climatic episode. Rather, the story is one of continual local and regional adjustment to fluctuating environmental, social, and demographic conditions. Instead of the fairly stable organizational configurations that should characterize adaptation to a distinct climatic regime, several major sociocultural transformations characterized this period. Thus, the MWP on the Colorado Plateau cannot be viewed as an interval of relatively consistent climate that promoted a narrow range of behavioral responses. Instead, it was a period of significant environmental variability that may have been modified to some degree by global processes that produced the MWP in Europe.

5. Summary and Conclusions

The survey of existing Colorado Plateau data for environmental or behavioral patterns that can be equated with the Medieval Warm Period is somewhat inconclusive. No prolonged environmental deviations corresponding to the entire 400-yr period of the MWP can be identified. Rather, considerable variability in low and high frequency environmental factors and in human adaptive behavior characterized the 900–1300 interval. Since variation of this sort is typical of the last 2000 yr, the MWP does not depart appreciably from the general regional pattern. On the other hand, it is possible that the global climate processes responsible for the MWP modified Southwestern conditions enough to leave traces in Plateau paleoenvironmental and behavioral data. Few such vestiges are evident in low frequency variability, which seems to reflect processes with internal dynamics not directly related to the climatic factors that produced the MWP. Future, higher resolution studies of low frequency environmental variability might, however, disclose spatially or temporally localized attributes that can be related to the MWP. Aspects of the high frequency record that may indicate MWP-related modifications of Plateau climate include the comparative rarity of intense droughts, the prevalence of low temporal variability, and major droughts in the middle 12th and late 13th centuries. Furthermore, the coherence exhibited by several low and high frequency environmental variables between 900 and 1200 may be a Southwestern effect of the MWP. In a similar fashion, human adaptive behavior between 900 and 1300 is highly varied and does not exhibit a general pattern that can be related to the MWP. Eventually it may be possible to identify behavioral adjustments to environmental fluctuations linked to the MWP; however, it currently is impossible to segregate such responses from those made to factors not related to the MWP.

Given these somewhat equivocal results, it may be asked whether the Colorado Plateau data advance understanding of the Medieval Warm Period. A partial answer to this question may be found in the implications of these data for the issues raised at the Tucson MWP conference (Diaz and Hughes, 1991). The absence of strong environmental and behavioral patterns during the 900–1300 interval cannot be taken as evidence that the MWP did not exist because, as the other papers in this issue demonstrate, relevant environmental excursions occur in Europe and elsewhere. The Plateau data do tend to place limits on the universality of the MWP by showing either that it was not a global phenomenon or that it was expressed differently in different regions. A few unusual attributes of Plateau environmental variability during the target interval may represent MWP-related modulations of the strong regional pattern of variation. The data also indicate that any MWP effects on the Colorado Plateau were not manifested over the whole 400-yr range of the MWP and, in fact, may have been restricted to the 900–1150 interval. Because the southern Colorado Plateau is a marginal area for farming, the archaeological consequences of a prolonged increase in temperature would be unmistakable. The absence of evidence for widespread subsistence failure or the sociocultural consequences thereof shows that the MWP does not represent a worldwide increase in temperature. The chief conclusion that can be drawn from the Plateau data is that the global climatic processes that produced the MWP had different effects in the Southwest than in northern Europe.

The foregoing survey of Plateau paleoenvironmental and archaeological data suggests several lines of further research to elucidate the nature and causes of the Medieval Warm Period. First, it would be extremely helpful to have an atmospheric system model relating climate conditions across the northern hemisphere. Such a model would allow climatic conditions on the Colorado Plateau to be predicted from the known attributes of the MWP in Europe. Such predictions would foster the development of testable hypotheses concerning both paleoenvironmental indicators of the MWP and human behavioral responses to environmental fluctuations related to the MWP. Second, the spatial and temporal manifestations of the MWP will have to be empirically identified in many regions of the world, a process that is well underway in many areas. Finally, the various local expressions will have to be synthesized within the context of the general model to understand the underlying atmospheric processes that produced the climatic excursion initially identified in northern Europe as the Medieval Warm Period.

Acknowledgements

The dendrochronological research underlying this paper was supported by the National Park Service, Southwestern Parks and Monuments Association, the Anthropology and Archaeometry Programs of the National Science Foundation, and the DOD Advanced Research Projects Agency. The discussion benefitted from the expertise of countless archaeologists and paleoenvironmentalists. Among these many individuals, I owe special thanks to B. Bannister, R. C. Euler, H. C. Fritts, D. A. Graybill, G. J. Gumerman, R. H. Hevly, T. N. V. Karlstrom, F. Plog, W. J. Robinson, and M. R. Rose. Two anonymous reviewers provided valuable suggestions for clarifying the manuscript. A. E. Dean produced Figure 1.

References

Adams, E. C.: 1991, *The Origin and Development of the Pueblo Katsina Cult.*, The University of Arizona Press, Tucson.

Baerreis, D. A., Bryson, R. A., and Kutzbach, J. E.: 1976, 'Climate and Culture in the Western Great Lakes Region', *Midcont. J. Archaeol.* **1**(1), 39–57.

Betancourt, J. L.: 1990, 'Late Quaternary Biogeography of the Colorado Plateau', in Betancourt, J. L., Van Devender, T. R., and Martin, P. S., (eds.), *Packrat Middens: The Last 40,000 Years of Biotic Change*, The University of Arizona Press, Tucson, pp. 259–292.

Betancourt, J. L. and Van Devender, T. R.: 1981, 'Holocene Vegetation in Chaco Canyon, New Mexico', *Science* **214**, 656–658.

Betancourt, J. L., Martin, P. S., and Van Devender, T. R.: 1983, 'Fossil Packrat Middens from Chaco Canyon, New Mexico: Cultural and Ecological Significance', in Wells, S. G., Love, D. W., and Gardner, T. W. (eds.), 'Chaco Canyon Country: A Field Guide to the Geomorphology, Quaternary Geology, Paleoecology, and Environmental Geology of Northwestern New Mexico', *American Geomorphological Field Group, 1983 Field Trip Guide Book*, Albuquerque, pp. 207–217.

Bohrer, V. L.: 1972, 'Paleoecology of the Hay Hollow Site, Arizona', *Fieldiana: Anthropology* **63**, 1–30, Field Museum of Natural History, Chicago.

Bryan, K.: 1954, 'The Geology of Chaco Canyon New Mexico in Relation to the Life and Remains of the Prehistoric Peoples of Pueblo Bonito', *Smiths. Misc. Coll.* **122**(7), Washington.

Buge, D. E. and Schoenwetter, J.: 1977, 'Pollen Studies at Chimney Rock Mesa', in Eddy, F. W. (ed.), *Archaeological Investigations at Chimney Rock Mesa: 1970–1972, Memoirs of the Colorado Archaeological Society*, No. 1, Boulder, pp. 77–80.

Burns, B. T.: 1983, *Simulated Anasazi Storage Behavior Using Crop Yields Reconstructed from Tree Rings: A.D. 652–1968*, Ph.D. Dissertation, Department of Anthropology, The University of Arizona, Tucson, University Microfilms International, Ann Arbor.

Butzer, K. W.: 1982, *Archaeology as Human Ecology: Method and Theory for a Contextual Approach*, Cambridge University Press, Cambridge.

Cooley, M. E.: 1962, 'Late Pleistocene and Recent Erosion and Alluviation in Parts of the Colorado River System, Arizona and Utah', 'Geological Survey Research 1962: Short Papers in Geology, Hydrology, and Topography, Articles 1–59', *United States Geological Survey Professional Paper* **450-B**, Washington, pp. 48–50.

Cordell, L. S.: 1984, *Prehistory of the Southwest*, Academic Press, Orlando.

Cordell, L. S. and Gumerman, G. J. (eds.): 1989, *Dynamics of Southwest Prehistory*, Smithsonian Institution Press, Washington.

Dean, J. S.: 1966, 'The Pueblo Abandonment of Tsegi Canyon, Northeastern Arizona', paper presented at the 31st Annual Meeting of the Society for American Archaeology, Reno.

Dean, J. S.: 1969, 'Chronological Analysis of Tsegi Phase Sites in Northeastern Arizona', *Pap. Labor. Tree-Ring Res.* **3**, The University of Arizona Press, Tucson.

Dean, J. S.: 1986, 'Paleoenvironment', manuscript on file at the Center for Archaeological Investigations, Southern Illinois University, Carbondale.

Dean, J. S.: 1987, 'Prehistoric Behavioral Response to Abrupt Environmental Changes in Northeastern Arizona', paper presented at the Conference on Civilization and Rapid Climatic Change, The Calgary Institute for the Humanities, The University of Calgary.

Dean, J. S.: 1988a, 'A Model of Anasazi Behavioral Adaptation', in Gumerman, G. J. (ed.), *The Anasazi in a Changing Environment*, Cambridge University Press, Cambridge, pp. 25–44.

Dean, J. S.: 1988b, 'Dendrochronology and Paleoenvironmental Reconstruction on the Colorado Plateaus', in Gumerman, G. J. (eds.), *The Anasazi in a Changing Environment*, Cambridge University Press, Cambridge, pp. 119–167.

Dean, J. S.: 1991, 'Social Complexity and Environmental Variability in Northeastern Arizona', paper presented at the 24th Chacmool Conference 'Culture and Environment: A Fragile Coexistence', The University of Calgary.

Dean, J. S.: 1992, 'Environmental Factors in the Evolution of the Chacoan Sociopolitical System', in Doyel, D. E. (ed.), *Anasazi Regional Organization and the Chaco System, Maxwell Museum of Anthropology, Anthropological Papers*, **5**, University of New Mexico, Albuquerque, pp. 35–43.

Dean, J. S. and Robinson, W. J.: 1977, *Dendroclimatic Variability in the American Southwest, A.D. 680–1970*, U.S. Department of Commerce, National Technical Information Service, PB-266 340, Springfield, Virginia.

Dean, J. S. and Robinson, W. J.: 1978, 'Expanded Tree-Ring Chronologies for the Southwestern United States', *Chronology Series III*, Laboratory of Tree-Ring Research, The University of Arizona, Tucson.

Dean, J. S., Euler, R. C., Gumerman, G. J., Plog, F., Hevly, R. H., and Karlstrom, T. N. V.: 1985, 'Human Behavior, Demography, and Paleoenvironment on the Colorado Plateaus', *Amer. Antiq.* **50**, 537–554.

Diaz, H. F. and Hughes, M. K.: 1991, 'Reconstruction of Spatial Patterns of Climatic Anomalies during the Medieval Warm Period (A.D. 900–1300): Interim Report on the Workshop Held in Tucson, Arizona, November 5–8 1991', manuscript on file at the Laboratory of Tree-Ring Research, The University of Arizona, Tucson.

Douglass, A. E.: 1929, 'The Secret of the Southwest Solved by Talkative Tree Rings', *Nation. Geogr. Magaz.* **56**, 736–770.

Douglass, A. E.: 1935, 'Dating Pueblo Bonito and Other Ruins of the Southwest', *Nation. Geogr. Soc. Contr. Techn. Pap., Pueblo Bonito Ser.* **1**, Washington.

Eddy, F. W.: 1964, 'The Alluvial Research', in Eddy, F. W. and Schoenwetter, J. (eds.), 'Alluvial and Palynological Reconstruction of Environments, Navajo Reservoir District', *Mus. New Mex. Pap. Anthropol.* **13**, Santa Fe, pp. 25–62.

Euler, R. C., Gumerman, G. J., Karlstrom, T. N. V., Dean, J. S., and Hevly, R. H.: 1979, 'The Colorado Plateaus: Cultural Dynamics and Paleoenvironment', *Science* **205**, 1089–1101.

Fall, P. L., Kelso, G., and Markgraf, V.: 1981, 'Paleoenvironmental Reconstruction at Canyon del Muerto, Arizona, Based on Principal-Component Analysis', *J. Archaeol. Sci.* **8**, 297–307.

Garza, P. L.: 1978, 'Retrodiction of Subsistence Stress on the Pajarito Plateau: A Computer-Based Model', Master's thesis, Department of Anthropology, University of California at Los Angeles.

Gregory, H. E.: 1916, 'The Navajo Country: A Geographic and Hydrographic Reconnaissance of Parts of Arizona, New Mexico, and Utah', *United States Geological Survey Water Supply Paper* **380**, Washington.

Gregory, H. E.: 1917, 'Geology of the Navajo Country: A Reconnaissance of Parts of Arizona, New Mexico, and Utah', *United States Geological Survey Professional Paper* 93, Washington.

Gumerman, G. J. (ed.): 1988, *The Anasazi in a Changing Environment*, Cambridge University Press, Cambridge.

Hack, J. T.: 1942, 'The Changing Physical Environment of the Hopi Indians of Arizona', 'Reports of the Awatovi Expedition, Peabody Museum, Harvard University', No. 1, *Papers of the Peabody Museum of American Archaeology and Ethnology, Harvard University*, Vol. 35, No. 1, Cambridge.

Hack, J. T.: 1945, 'Recent Geology of the Tsegi Canyon', in Beals, R. L., Brainerd, G. W., and Smith, W. (eds.), *Archaeological Studies in Northeast Arizona, University of California Publications in American Archaeology and Ethnology*, Vol. 44, No. 1, University of California Press, Berkeley and Los Angeles, pp. 151–158.

Hall, S. A.: 1977, 'Late Quaternary Sedimentation and Paleoecologic History of Chaco Canyon, New Mexico', *Geol. Soc. Amer. Bull.* **88**, 1593–1618.

Hall, S. A.: 1983, 'Holocene Stratigraphy and Paleoecology of Chaco Canyon', in Wells, S. G., Love, D. W., and Gardner, T. W. (eds.), *Chaco Canyon Country: A Field Guide to the Geomorphology, Quaternary Geology, Paleoecology, and Environmental Geology of Northwestern New Mexico, American Geomorphological Field Group, 1983 Field Trip Guidebook*, Albuquerque, pp. 219–226.

Haury, E. W.: 1935, 'Tree-Rings – the Archaeologist's Time-Piece', *Amer. Antiq.* **1**, 98–108.

Hevly, R. H.: 1964, 'Paleoecology of Laguna Salada', in Martin, P. S., Rinaldo, J. B., Longacre, W. A., Freeman, Jr., L. G., Brown, J. A., Hevly, R. H., and Cooley, M. E., *Chapters in the Prehistory of Eastern Arizona, II, Fieldiana: Anthropology* **55**, Chicago Natural History Museum, Chicago, pp. 171–187.

Hevly, R. H.: 1981, 'Pollen Production, Transport, and Preservation: Potentials and Limitations in Archaeological Palynology', *J. Ethnobiol.* **1**(1), 39–54.

Hevly, R. H.: 1988, 'Prehistoric Vegetation and Paleoclimates on the Colorado Plateaus', in Gumerman, G. J. (ed.), *The Anasazi in a Changing Environment*, Cambridge University Press, Cambridge, pp. 92–118.

Jorde, L. B.: 1977, 'Precipitation Cycles and Cultural Buffering in the Prehistoric Southwest', in Binford, L. R. (ed.), *For Theory Building in Archaeology*, Academic Press, New York, pp. 385–396.

Karlstrom, E. T.: 1983, 'Soils and Geomorphology of Northern Black Mesa', in Smiley, F. E., Nichols, D. L., and Andrews, P. P. (eds.), *Excavations on Black Mesa, 1981: A Descriptive Report, Center for Archaeological Investigations Research Paper*, No. 36, Southern Illinois University, Carbondale, pp. 317–342.

Karlstrom, E. T. and Karlstrom, T. N. V.: 1986, 'Late Quaternary Alluvial Stratigraphy and Soils of the Black Mesa – Little Colorado River Areas, Northern Arizona', in Nations, J. D., Conway, C. M., and Swann, G. A. (eds.), *Geology of Central and Northern Arizona: Geological Society of America, Rocky Mountain Section, Guidebook*, Flagstaff, pp. 71–92.

Karlstrom, T. N. V.: 1988, 'Alluvial Chronology and Hydrologic Change of Black Mesa and Nearby Regions', in Gumerman, G. J. (ed.), *The Anasazi in a Changing Environment*, Cambridge University Press, Cambridge, pp. 45–91.

Karlstrom, T. N. V., Gumerman, G. J., and Euler, R. C.: 1976, 'Paleoenvironmental and Cultural Correlates in the Black Mesa Region', in Gumerman, G. J. and Euler, R. C. (eds.), *Papers on the Archaeology of Black Mesa, Arizona*, Southern Illinois University Press, Carbondale, pp. 149–161.

Kelso, G. K.: 1976, *Absolute Pollen Frequencies Applied to the Interpretation of Human Activities in Northern Arizona*, Ph.D Dissertation, The University of Arizona, Tucson, University Microfilms International, Ann Arbor.

Lamb, H. H.: 1977, *Climate: Present, Past and Future, Vol. 2, Climate History and the Future*, Methuen, London.

Lebo, C. J.: 1991, *Anasazi Harvests: Agroclimate, Harvest Variability, and Agricultural Strategies on Prehistoric Black Mesa, Northeastern Arizona*, Ph.D Dissertation, Indiana University, Bloomington, University Microfilms International, Ann Arbor.

Lipe, W. D. and Matson, R. G.: 1975, 'Archaeology and Alluvium in the Grand Gulch – Cedar Mesa Area, Southeastern Utah', in *Four Corners Geological Society Guidebook, Eighth Field Conference: Canyonlands*, pp. 69–71.

Love, D. W.: 1980, *Quaternary Geology of Chaco Canyon, Northwestern New Mexico*, Ph.D Dissertation, Department of Geology, University of New Mexico, Albuquerque, University Microfilms International, Ann Arbor.

Petersen, K. L.: 1988, 'Climate and the Dolores River Anasazi', *Univ. Utah Anthropol. Pap.* **113**, University of Utah Press, Salt Lake City.

Plog, F., Gumerman, G. J., Euler, R. C., Dean, J. S., Hevly, R. H., and Karlstrom, T. N. V.: 1988, 'Anasazi Adaptive Strategies: The Model, Predictions, and Results', in Gumerman, G. J. (ed.), *The Anasazi in a Changing Environment*, Cambridge University Press, Cambridge, pp. 230–276.

Rose, M. R., Dean, J. S., and Robinson, W. J.: 1981, 'The Past Climate of Arroyo Hondo, New Mexico, Reconstructed from Tree Rings', *Arroyo Hondo Archaeol. Ser.* **4**, School of American Research Press, Santa Fe.

Rose, M. R., Robinson, W. J., and Dean, J. S.: 1982, 'Dendroclimatic Reconstruction for the Southeastern Colorado Plateau', manuscript on file at the Laboratory of Tree-Ring Research, The University of Arizona, Tucson.

Schoenwetter, J.: 1962, 'The Pollen Analysis of Eighteen Archaeological Sites in Arizona and New Mexico', in Martin, P. S., Rinaldo, J. B., Longacre, W. A., Cronin, C., Freeman, Jr., L. G., and Schoenwetter, J., *Chapters in the Prehistory of Eastern Arizona, I, Fieldiana: Anthropology*, **53**, Chicago Natural History Museum, Chicago, pp. 168–209.

Schoenwetter, J.: 1966, 'A Re-Evaluation of the Navajo Reservoir Pollen Chronology', *El Palacio* **73**(1), pp. 19–26.

Schoenwetter, J.: 1967, 'Pollen Survey of the Chuska Valley', in Harris, A. H., Schoenwetter, J., and Warren, A. H., *An Archaeological Survey of the Chuska Valley and the Chaco Plateau, New Mexico, Part I, Natural Sciences, Museum of New Mexico Research Records* **4**, Museum of New

Mexico Press, Santa Fe, pp. 72–103.

Schoenwetter, J.: 1970, 'Archaeological Pollen Studies of the Colorado Plateau', *Amer. Antiq.* **35**, pp. 35–48.

Sears, P. B.: 1937, 'Pollen Analysis as an Aid in Dating Cultural Deposits in the United States', in MacCurdy, G. G. (ed.), *Early Man as Depicted by Leading Authorities at the International Symposium, The Academy of Natural Sciences, Philadelphia, March 1937*, J. B. Lippencott Company, London, pp. 61–66.

Thornthwaite, C. W., Sharpe, C. F. S., and Dosch, E. F.: 1942, 'Climate and Accelerated Erosion in the Arid and Semi-Arid Southwest, with Special Reference to the Polacca Wash Drainage Basin, Arizona', *United States Department of Agriculture Technical Bulletin*, No. 808, Washington.

Van West, C. R.: 1990, *Modeling Prehistoric Climatic Variability and Agricultural Production in Southwestern Colorado: A GIS Approach*, Ph.D Dissertation, Department of Anthropology, Washington State University, Pullman, University Microfilms International, Ann Arbor.

Vivian, R. G.: 1990, *The Chacoan Prehistory of the San Juan Basin*, Academic Press, San Diego.

(Received 26 January, 1993; in revised form 11 October, 1993)

A WARM AND WET LITTLE CLIMATIC OPTIMUM AND A COLD AND DRY LITTLE ICE AGE IN THE SOUTHERN ROCKY MOUNTAINS, U.S.A.*

KENNETH LEE PETERSEN

Westinghouse Hanford Company, P.O. Box 1970, Mail Stop H4-14, Richland, WA 99352, U.S.A.

Abstract. The zenith of Anasazi Pueblo Indian occupation in the northern Colorado Plateau region of the southwestern U.S.A. coincides with the Little Climatic Optimum or Medieval Warm Period (A.D. 900–1300), and its demise coincides with the commencement of the Little Ice Age. Indexes of winter (jet-stream derived) and summer (monsoon derived) precipitation and growing season length were developed for the La Plata Mountains region of southwestern Colorado. The results show that during the height of the Little Climatic Optimum (A.D. 1000–1100) the region was characterized by a relatively long growing season and by a potential dry farming zone or elevational belt (currently located between 2,000 m and 2,300 m elevation) that was twice as wide as present and could support Anasazi upland dry farming down to at least 1,600 m, an elevation that is quite impossible to dry farm today because of insufficient soil moisture. This expanded dry-farm belt is attributable to a more vigorous circulation regime characterized by both greater winter and summer precipitation than that of today. Between A.D. 1100 and 1300 the potential dry-farm belt narrowed and finally disappeared with the onset of a period of markedly colder and drier conditions than currently exist. Finally, when the Little Ice Age terminated in the mid A.D. 1800s and warmer, wetter conditions returned to the region, another group of farmers (modern Anglos) were able to dry farm the area.

1. Introduction

According to the greenhouse theory, during the next decades and centuries global climatic variation will exceed the historical records as the lower atmosphere warms in response to a rise in concentrations of carbon dioxide, methane, and other gases (Houghton *et al.*, 1990). The sharp contrast between the large predicted future change and the small climatic changes of the last century indicates that this latter period may offer an insufficient basis for appreciating the projected future climate and vegetation changes. Examination of larger-than-historic climatic changes that have occurred in the past, in specific locations, may provide a context for evaluating possible future changes (Schneider, 1986). Estimates of regional climate cannot be viewed in isolation but must be viewed as part of a larger continental and global system to understand fully the underlying driving mechanisms.

Anyone visiting the prehistoric ruins in the American Southwest leaves with a sense of respect for the ancient dwellers, the construction of their dwellings, and

* The U.S. Government right to retain a non-exclusive, royalty-free license in and to any copyright is acknowledged.

the balance they achieved with nature. The latter aspect is particularly interesting because these ancient people relied so heavily on agriculture in an area recognized for its arid climate and, in some regions, relatively high elevation. The earliest populations before the time of Christ were originally hunters and gatherers, but during the last 2,000 yr they have evolved into three culturally distinct traditions upon stimulus from Mexico. These include the Mogollon or Western Pueblo, the more widespread Anasazi Pueblo of the northern Southwest, and the regionally restricted Hohokam culture, which was confined to the Salt and Gila River drainages of southern Arizona. By A.D. 900 (Figure 1), these cultural traditions had developed advanced and flourishing societies, farming corn, beans, and squash, and supplementing their diet by hunting and gathering or raising food, such as turkeys.

By A.D. 1100 the Anasazi (which, in this case, also includes the corn-growing Fremont and Sevier-Fremont groups of present-day Utah) had reached their most northern extension. Around A.D. 1200, long before any Europeans – or even the Navajo or Apache Indians – arrived in the region, the Anasazi began to abandon most of their former northern territory (currently Colorado and Utah). At the same time, the population in the Rio Grande Valley began to expand rapidly, probably from immigration, and it continued to grow until its peak at about A.D. 1300. Other Anasazi groups migrated to parts of Arizona. By A.D. 1300, the area of Utah and Colorado no longer had evidence of Indian farmers growing corn, suggesting the region had been vacated and during the succeeding several centuries the population of the remaining territory was reduced as other Indian groups and Europeans moved into the southwest United States. However, the Anasazi did not disappear; they probably became the Hopi, Zuni, and other modern Pueblo Indians of northern Arizona and New Mexico, some of which still exist today.

The zenith of Anasazi Pueblo Indian occupation coincides with the Little Climatic Optimum (Medieval Warm Period) of Europe (A.D. 900–1300), a time of purported elevated temperatures compared to those of the present (Lamb, 1977). The exact nature of the climate and vegetation manifested in the Four Corners region (where Utah, Colorado, Arizona, and New Mexico meet) during that time has received only cursory quantification. However, the precarious nature of farming in the Four Corners region has led many researchers to hypothesize that any severe climatic deterioration in the past may have affected the Anasazi's ability to grow corn the same way that it could affect efforts today (see Gumerman, 1988, and references therein).

During recent decades, many scietists have maintained, on the basis of tree-ring evidence, that drought forced the Anasazi from the Four Corners region. However, other scientists refer to indications that there had been earlier droughts that were as serious, yet the Anasazi did not leave during those droughts (Fritts et al., 1965). Some researchers (e.g., Berlin et al., 1977; Bray, 1971; Bryson and Julian, 1963; Martin and Byers, 1965; Smiley, 1961; Woodbury, 1961) have suggested that the demise of the Anasazi culture may coincide with the onset of the Little Ice Age, a time of colder temperatures than the present (Grove, 1988). In many parts of the

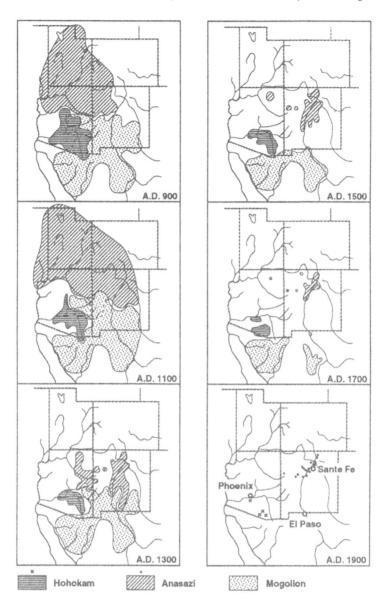

Fig. 1.   Map showing the approximate event of Southwest United States Indian cultures at two-hundred-year intervals (redrawn from Jennings, 1968, Figure 7.2.).

world, the Little Ice Age has been described as a time of renewed glacial activity, expanding snowfields and, in some regions, reduced summer monsoons. However, the evidence for the timing, severity, and exact nature of that climatic episode in the Four Corners region generally is unknown.

Presented here is a specific case study of the Anasazi Pueblo Indians' farmers of the Four Corners region of the southwest United States, focusing on the timing,

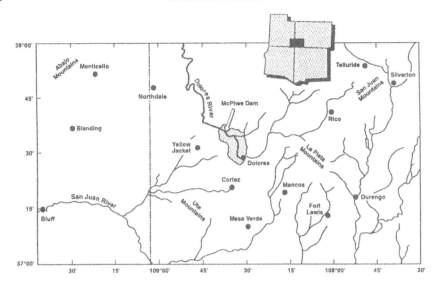

Fig. 2. Map showing the location of the mountains, the Dolores Project site (shaded), McPhee Dam, and area weather stations in southwestern Colorado and adjacent Utah listed in Table I.

nature, and range of climatic and vegetation change coincident with the Anasazi occupation and abandonment of that high plateau region. The research summarized here began in 1972 in conjunction with the Salmon Ruins Archaeological Project (northeast New Mexico) and continued as part of the archaeological mitigation effort (1978–1985) necessitated by the U.S. Bureau of Reclamation's construction of the McPhee Dam and Reservoir on the Dolores River in southwest Colorado (Breternitz *et al.*, 1986; Petersen and Mehringer, 1976; Petersen *et al.*, 1985; Petersen *et al.*, 1987; Petersen, 1988a).

2. The La Plata Mountains Regional Environment

The La Plata Mountains of southwestern Colorado are a remote and picturesque mountain group that protrudes into the eastern edge of the Colorado Plateau 30 km southwest of the main San Juan Mountain front. Contrasting the adjacent relatively arid plateau, the La Plata Mountains and the intervening rugged hill region between them and the San Juan Mountains are well watered and well timbered. Several peaks in the La Plata exceed 3,660 m. The north side of the La Plata Mountains drains into the Dolores River, and the remaining sides flow into the San Juan River (Figure 2).

Although climatic data are not available for the La Plata Mountains specifically, Bradley and Barry (1973) found that precipitation records from stations throughout southwestern Colorado were highly correlated. Climate records (Table I) illustrate increased precipitation and reduced temperature with elevation. Barry and Bradley (1976) indicate a summer lapse rate for the San Juan National Forest in southwest Colorado of –0.82 °C/100 m and an annual lapse rate of –0.59 °C/100 m, while

TABLE I: Climatic Summary (1951–1980) for 14 Weather Stations in Southwestern Colorado and Southeastern Utah Shown in Figure 2

Station Name	Elevation (m)	Annual Precipitation (cm)	Mean January Temperature (°C)	Mean July Temperature (°C)
Silverton, CO	2,842	56.59	–8.6	13.3
Rico, Co	2,695	66.57	#	#
Telluride, CO	2,669	54.79	–5.7	15.4
Fort Lewis, CO	2,315	44.78	–5.4	18.1
Mesa Verde, CO	2,155	44.55	–1.6	22.6
Mancos, CO	2,140	40.51	#	#
Monticello, UT	2,128	36.60	–3.9	20.3
Dolores, CO	2,118	45.90	#	#
Yellow Jacket 2W, CO*	2,091	37.80	–4.6	21.4
Northdale, CO	2,040	30.58	–5.2	20.3
Durango, CO	1,996	47.27	–3.5	20.2
Cortez, CO	1,883	31.17	–2.8	22.2
Blanding, UT	1,868	29.72	–2.6	23.1
Bluff, UT	1,315	19.33	–1.3	25.6

\# Station records precipitation only.
* For years 1962–1979.

Betancourt (1984) reports a slightly lower annual figure for southeastern Utah (–0.45 °C/100 m).

The precipitation distribution through the year in the Four Corners region is bimodal (Figure 3), with pronounced cool and warm season maxima, the latter of which has been recognized as a monsoon (Bryson and Lowry, 1955; Huntington, 1914). The monsoon precipitation in the Four Corners region results primarily from moisture sweeping from the south (Figure 4). South and southeast of the monsoon boundary (and mostly at higher elevations where the growing season is still adequate), the modern dry farmers raise summer crops such as corn, beans, and potatoes. North and west of that monsoon boundary at higher elevation, the soil moisture obtained from winter precipitation is usually adequate to allow dry farmers to raise winter and spring wheat. However, by midsummer soil moisture must be supplemented with irrigation to allow the summer crops to mature.

The climatic differences in the 2,300 m of elevational change between the San Juan River and the highest La Plata Mountain peaks provide several vegetation zones. Figure 5 reconstructs the native vegetation circa 1920, which was just before the extensive use of the tractor. The vegetation reconstruction is based on old photographs, historic descriptions, maps, and fossil pollen evidence (Petersen *et al.*, 1985; Petersen *et al.*, 1987).

The extensive stands of sagebrush (*Artemisia tridentata*) that occur to the west

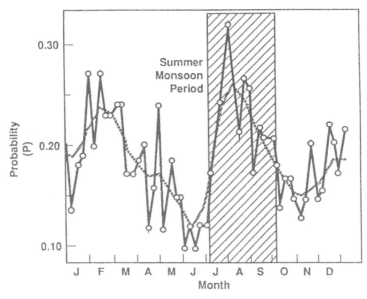

Fig. 3.   Probability at Cortez, Coloraado (1931–1961) that a given day of the year would be wet (≥ 0.25 mm precipitation). Heermann *et al.* (1971) computed daily average probabilities, then transformed these into weekly averages. Plots are on the midpoint of each 7-day period. Smoothing (dotted line) was done using a 9-level, weighted moving binomial running mean (e.g., $T_0' = [1T_{-4} + 2T_{-3} + 4T_{-2} + 6T_{-1} + 8T_{-0} + 6T_{+1} ...]/34$), and hatching shows the period defined here as the summer monsoon.

of the La Plata Mountains led Newberry (1876, p. 84) to name the divide between the San Juan and Dolores rivers (extending from Mesa Verde west to Comb Ridge, near Blanding, Utah) the 'Sage Plain'. This sagebrush-covered plateau (10,360 km^2) ranges between 1,500 and 2,100 m elevation (Gregory and Thorpe, 1938). Much of the area today between 2,010 m and 2,380 m elevation has been cleared of natural vegetation and is under dry-land cultivation. In Figure 5 that zone (or elevational belt) of cultivation represents primarily the higher elevations of the pinyon and juniper, most of the big sagebrush, and the lower elevational limit of the montane scrub mapping units. This relatively narrow agricultural belt is farmable because of the good soils and because it is both wet enough (greater than 35.5 cm of annual precipitation – of which 10 cm fall during the warm season) and warm enough (including a greater-than-110-day frost-free season) to allow routine dry farming of such crops as corn, beans, potatoes, and grains (Petersen *et al.*, 1987; U.S. Department of Agriculture, Soil Conservation Service, 1976).

When farming within this belt, contemporary farmers often discover the cultural remains of ancient Anasazi, who farmed corn, beans, and squash on the same land hundreds of years earlier. The flanks of the La Plata Mountains in southwestern Colorado are surrounded by vast areas containing evidence of Anasazi occupation, the most famous being preserved in the boundaries of the Mesa Verde National

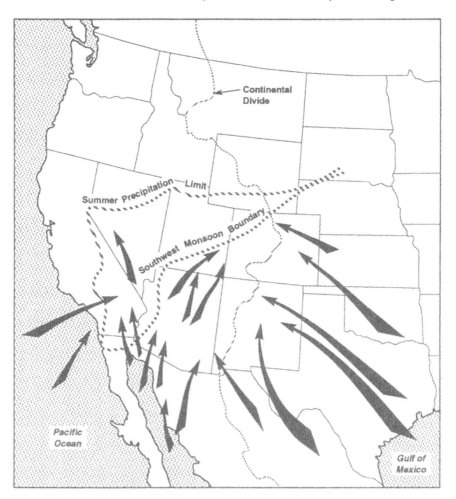

Fig. 4. Climatic boundaries for the Southwest monsoon. Precipitation is greatest east and south of the southwest monsoon boundary of Mitchell (1976), where more than half of the annual precipitation occurs during the warm season (Dorrah, 1946). North of that boundary the amount of warm season precipitation decreases until it reaches the summer precipitation limit of Pyke (1972). Arrows show the main paths of moisture in the southwest United States during the summer (adapted from Miller *et al.*, 1973)

Park (Osborne, 1965; Wormington, 1947; Cordell, 1984). Because modern changes in either moisture or summer warmth could affect the width or elevational range of the present potential dry-farming belt, this must also have been true for the Anasazi.

3. Research Strategy

In this study, the prehistoric elevation and width of the dry farming belt is reconstructed and contrasted with that of the present, based on the premise that

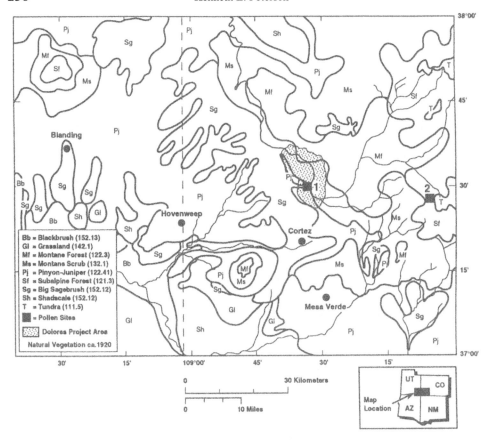

Fig. 5. Major plant communities in southwestern Colorado and southeastern Utah, ca. 1920 (Petersen *et al.*, 1987). Numbers in parentheses refer to Brown's (1982) vegetation classification scheme. Black square number 1 shows the location of Sagehen Marsh. Black square number 2 shows the location of the Beef Pasture and Twin Lakes pollen-coring sites; they differ by 230 m in elevation and are located only 5 km apart.

knowledge of the history of that belt would be useful in unveiling the history of Anasazi settlement and movement. This is because the climatic factors that affect horticulture (i.e., sufficient growing season and moisture for dry farming) are generally the same factors affecting the natural plant community distribution and composition. The index used for hypothesizing the width of the dry-farming belt is the changing width and elevation of the spruce (*Picea engelmannii*) forest zone in the La Plata Mountains. This index was obtained by analyzing pollen records from two different elevations within the spruce zone. The lower-elevation record provided a history of the moisture-dependent lower-elevation spruce boundary while the upper-elevation record provided a history of the temperature-dependent timberline. Because spruce growth responds differently to climate variations at different elevations, the combined radiocarbon-dated pollen records from the two sites yield climate information not obtainable from either site alone. In addition, the changing record of pinyon (*Pinus edulis*) pollen, which wafted up from the surrounding lowlands and was deposited at the lower pollen site, is used as a measure of the changing number of pinyon trees on the landscape. A third pollen site that occurs within the potential dry-farm belt was used to verify some of the reconstructions obtained from the pollen records in the spruce forest. Some of the more important findings are summarized here; additional detail is reported elsewhere (Petersen *et al.*, 1985; Petersen *et al.*, 1987; Petersen, 1988a).

4. Results and Discussion

Location of Pollen Sites and the Spruce Forest Zone. Twin Lakes and Beef Pasture are located on the west slope of the La Plata Mountains, 45 km east of Cortez, Colorado, near the drainage divide between the Dolores and San Juan rivers (Figure 5). Twin Lakes (3290 m; NE1/4 SE1/4 NE1/4, sec. 18, T37N, R11W, La Plata, Colorado, 7.5 minute quadrangle) is 250 m below timberline in a depression near the headward snout of a landslide (Petersen and Mehringer, 1976). Twin Lakes and the adjacent sedge meadow are surrounded by an open Engelmann spruce and subalpine fir (*Abies lasiocarpa*) forest that has been logged recently.

Beef Pasture (3,060 m; S1/2 SW1/4, sec. 11, T37N, R12W, Rampart Hills, Colorado, 7.5 minute quadrangle) is an open 75-ha grass and sedge meadow surrounded by a mixed-conifer forest at the lower limit of the dense spruce-fir forest. The surrounding trees are spruce, Douglas-fir (*Pseudotsuga menziesii*), and aspen (*Populus tremuloides*).

The next lower vegetation zone in the La Plata Mountains is the Montane Forest (Figure 5), which has Ponderosa pine (*Pinus ponderosa*) and oak (*Querus gambeii*). Ponderosa is generally limited to between 2,250 and 2,750 m in elevation.

Confirmation of some aspects of the La Plata Mountain paleoenvironmental reconstruction are provided by a third pollen site, Sagehen Marsh (2,085 m; SW1/4 NE14; sec 36, T38N, R16W, Trible Point, Colorado, 7.5 minute quadrangle), located in the Dolores Project area in a valley that drains into the Dolores River (Petersen *et al.*, 1985; Petersen, 1985). Sagehen Marsh was dammed by an alluvial fan and surrounded by a mosaic of sagebrush, pinyon (*Pinus edulis*), juniper (*Juniperus osteosperma* and *Juniperus scopulorum*), and oak before being submersed in the McPhee Reservoir.

The stippled area in Figure 6 indicates the zone of relatively dense spruce around Twin Lakes in the La Plata Mountains. Results of surface pollen studies and documented historic changes indicate that the large, heavy spruce pollen does not travel very far from the source tree (King, 1967; Maher, 1963; Wright, 1952) and so is a good indicator of the proximity of the trees. The location of the two pollen sites makes them especially sensitive to the changes in the width of the spruce forest. A depression of upper timberline of only 100 m would decrease the area occupied by forest nearly 25 km^2, doubling the area above timberline. Actually, the pollen record at Twin Lakes does respond to changes in elevation of upper timberline,

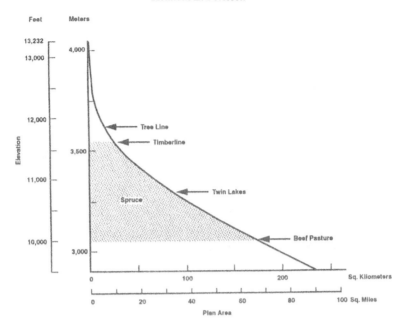

Fig. 6. Area elevation graph for 250 km^2 around Twin Lakes, La Plata Mountains, Colorado. The curve is drawn from a plot of the plan area above each 152-m contour (La Plata, Colorado, 7.5 minute quadrangle), beginning at 2895 m.

especially to timberline receding to lower elevations. Beef Pasture's location (at the current lower elevational limit of dense spruce forest) makes it particularly sensitive to the elevational changes in that boundary.

Dating and Pollen Zones. Age determination of vegetation change in the La Plata Mountains is aided by 17 radiocarbon dates (Figure 7). Because Beef Pasture accumulated almost twice as much sediment over the last 5,600 yr, the depth scales in Figure 7 have been adjusted to allow clear comparison with Twin Lakes. The lines between samples are approximations based on visual judgement. This approach (instead of using a linear regression) was used because of the apparent correspondence in the depositional histories between the two sites. The sediments from both sites are primarily peats that have been shown, in some environments, to have varying rates of deposition depending on the prevailing climate. Slower accumulation rates usually occur during period of shorter, colder, drier summers, while faster rates occur during longer, warmer, wetter summers (Malaurie *et al.*, 1972, p. 116; Nichols, 1975, pp. 60–70; Short and Nichols, 1977, p. 288). Examination of the deposition rate curves for both Twin Lakes and Beef Pasture between 5,000 and 2,500 yr B.P. show marked similarities. For instance, the deposition rate between 3,000 and 2,500 yr B.P. increased at each site, and the comparable age estimates for the zone 3/4 boundary (approximately 3,400 yr B.P.) and the zone 4/5 boundary (approximately 2,700 yr B.P.) are relatively constrained by radiocarbon dates.

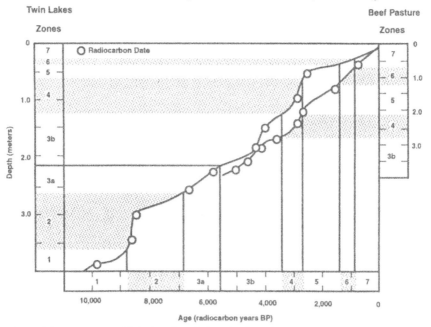

Fig. 7. Deposition-rate curves (fit by hand) and plot of radiocarbon sample for Twin Lakes and Beef Pasture, and the correlation of pollen zones shown by stippling. Note different depth scales. A full description of radiocarbon dates is found in Petersen (1988a) and Petersen and Mehringer (1976); see Table II for those in this study.

A computer clustering program was used to collect adjacent pollen samples into groups that were most similar. The assumption is that a climate change long and severe enough to affect vegetation in the La Plata Mountains would be reflected at a site by the clustering of samples into another group (Petersen, 1988a). The grouped samples were collected into a number of pollen zones, and the zones were numbered from bottom to top (Figure 7). Zone boundaries for each site were drawn to fall precisely on a single pollen sample so that the correlation of zone boundaries between the two pollen sites would be unambiguous. The radiocarbon age of the sample falling on the zone boundary was estimated as shown in Figure 7. Zone boundaries important to this study include those for zones 4/5, 5/6, and 6/7.

To facilitate correlation between dated regional Anasazi archeological sites and the La Plata Mountain pollen sequence the pollen sample ages were converted from the estimated radiocarbon ages derived from Figure 7 to tree-ring calibrated calendar dates. That conversion is discussed in more detail in the following sections. The results are believed to be satisfactory, but the imprecision of radiocarbon dates is clearly a weak link in a study that demands high-resolution dating.

Pollen Ratios and Their Historic Climatic Calibration. The interpretation of vegetation change presented here relies heavily on specific pollen ratios from Twin Lakes and Beef Pastures. At Beef Pasture, near the lower spruce-forest border, the

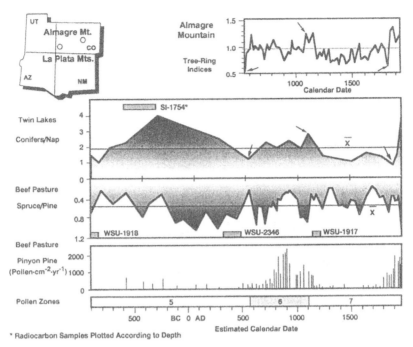

Fig. 8. Pollen ratios from the Twin Lakes and Beef Pasture pollen sites in the La Plata Mountains
drawn so that they emphasize the relative width of the spruce forest through time as compared to
a record of high-elevation bristlecone pine tree-ring indices from central Colorado (LaMarche and
Stockton, 1974) and estimate of the number of pinyon pollen grains deposited during 1 yr at Beef
Pasture. Arrows indicate points of correlation.

spruce to pine pollen-ratio curve was selected as the best record of the movement of
the lower limit of the spruce forest. Pinyon and ponderosa pine trees are the major
contributors to the pine portion of this ratio, and are trees that grow at elevations
below the spruce zone. To visually display the relative elevational changes in the
lower spruce border (Figure 8) the ratio value is plotted on a negative y-axis that
starts at zero and increases downward. Thus, increases in the spruce/pine ratio at
Beef Pasture are plotted farther from the origin and represent, relatively, a lowered
elevation limit for the lower spruce zone boundary.

Repeat photographs in the La Plata Mountains and surrounding regions document historic thickening (i.e., increase in density) and expansion to lower elevations
of the lower spruce forest border over the last 100 yr (Petersen, 1988a). This expansion has occurred despite the impact of logging and/or grazing. This forest
expansion probably corresponds to a concomitant increase in effective moisture
early this century, primarily in the form of increased winter or spring snow pack
the highest in the last 400 yr (Stockton, 1976; Thomas, 1959). Spruce is relatively
shallow rooted and cannot tolerate much soil drought, so a decrease in effective
moisture would be detrimental to tree survival, while an increase in effective moisture (especially snow) should foster growth (Dix and Richards, 1976; Daubenmire,

1943; Pearson, 1931).

At Twin Lakes, the proportion of conifer pollen (spruce + fir + pine) to pollen from nonarboreal plants (NAP) is selected to represent the relative fluctuations of the timberline (Figure 8) as it has been done in Europe (e.g., Patzelt, 1974). This ratio, rather than that used at Beef Pasture, was chosen because the Twin Lakes area recently was logged, thus artificially removing some spruce trees and probably affecting the modern Twin Lakes spruce-to-pine ratio. The ratio of all conifers to NAP was used to offset the effect of modern logging because less weight is given specifically to spruce in the ratio.

Repeat photographs document that the tree-line in the La Plata Mountains has risen about 50 m in the last 100 yr, and there has been a concomitant change in tree-growth form at the tree-line where krummholz form has changed to non-krummholz growth form. Rings from trees at timberline have changed from narrow to wide, coinciding with the extension of the growing season at the lower elevation during part of the same time period (Petersen, 1988a). Because the major limiting climatic factor for tree growth near timberline is summer temperature (Daubenmire, 1954; Tranquillini, 1979; Wardle, 1974), these changes indicate an increase in summer temperature from last century to this century.

The present pollen ratios at the modern surface of Twin Lakes and Beef Pasture (Figure 8) may be too high. Thousands of hectares of sagebrush and pinyon were cleared by modern farmers in the region after 1920 (Petersen *et al.*, 1987). Because these plants are wind pollinated, their pollen is widely distributed beyond the source plants. Extensive removal (such as the clearance by farmers) of these pollen sources would tend to decrease their proportion in the pollen rain deposited in the spruce forest and probably would enhance the relative proportion of spruce pollen deposited. The historic expansion at both the upper and lower limit of the spruce forest in the La Plata Mountains is documented, but the magnitude of that expansion as reflected in the modern surface pollen ratios at Twin Lakes and Beef Pasture (Figure 8) probably is exaggerated because of the bias introduced by the lowland clearing. The unbiased ratio should be below that of the modern but higher than the mean lines shown in Figure 8. At Beef Pasture, the ratio values of ≥ 0.60 are considered representative of conditions similar to those that have occurred historically, and values lower than these are considered drier.

Correlation and Dating of Twin Lakes and Beef Pasture Records. The age assignment given to each of the pollen samples in Figure 8 depends on many critical assumptions. One of these is that pollen zone boundaries can be correlated between the two sites. Zone 4/5 boundary (at the left edge of Figure 8) was assigned using radiocarbon dating, the results of the pollen sample clustering computer program, and the pollen ratio signature just to the right that indicates a relatively short period of time, characterized by a very narrow spruce forest. Because the recovery of the forest seems to be as rapid as the onset, the trees located near the climatically stressed upper and lower margin of the spruce forest zone simply may have cur-

TABLE II: Radiocarbon Dates Presented in this Study

Beef Pasture Radiocarbon Dates				
Depth (m)	Labora-tory Number	Date (yr B.P.)	Calibrated A.D./ B.C. Age, 1 SD, min (cal age) max*	Calibrated A.D./ B.C. Age, 1 SD, min (cal age) max**
0.500–0.550	WSU–1917	780±90	A.D. 1075 (1170) 1264	A.D. 1169 (1259) 1281
1.175–1.275	WSU–2346	1,540±80	A.D. 327 (410) 493	A.D. 420 (539) 605
2.000–2.050	WSU–1918	2,680±80	B.C. 812 (730) 647	B.C. 910 (828) 801

Twin Lakes Radiocarbon Date				
Depth (m)	Labora-tory Number	Date (yr B.P.)	Calibrated A.D./ B.C. Age, 1 SD, min (cal age) max*	Calibrated A.D./ B.C. Age, 1 SD, min (cal age) max**
0.40–0.49	SI–1754	2545±75	B.C. 685 (595) 505	B.C. 805 (790) 534

Sagehen Marsh Radiocarbon Date				
Depth (m)	Labora-tory Number	Date (yr B.P.)	Calibrated A.D./ B.C. Age, 2 SD, min – max***	Calibrated A.D./ B.C. Age, 1 SD, min (cal age) max**
1.42–1.61	Beta–3058	1350±60	A.D. 585 – 785	A.D. 643 (661) 682

*   Damon *et al.* (1974).
**   Stuiver and Reimer (1986).
*** Klein *et al.* (1982).

tailed pollen production temporarily, recovering when conditions improved (Hevly, 1981; Nichols, 1975, pp. 28–29) rather than reflecting only the loss of trees, which were later replaced by seedlings that grew to mature pollen producing trees. Another such narrow forest episode, along with the computer pollen clustering results, was used to correlate the samples on the zone 5/6 boundary. Finally, the correlation for zone 6/7 boundary was accomplished by aligning the samples showing the broadest spruce forest zone down core from the present.

None of these three anchor points have radiocarbon dates that coincide directly with them, but each has a date closely associated with it. Initially, the tables of Damon *et al.* (1974) were used to convert the radiocarbon age of the radiocarbon sample to a tree-ring corrected calendar age. Because the bracketing range of the standard deviation on the radiocarbon dates (Table II) is in excess of 100 yr, the calendar dates for the zone boundaries were not assigned in increments of less than a half century. The radiocarbon dates from both Twin Lakes and Beef Pasture were used to constrain the earliest narrow spruce zone episode shown on the extreme left of Figure 8; it was assigned an age of 800 B.C. The zone 4/5 boundary that falls on the left edge of the figure was assigned a calendar age of 900 B.C.; A.D. 550 was assigned to the zone 5/6 boundary; and A.D. 1100 was assigned to the widest spruce forest of zone 6 at the zone 6/7 boundary.

Once the calendar ages were assigned to the zone boundaries, the deposition curve based on those in Figure 7 was used to assign appropriate calendar ages to every sample using their intersection with the deposition rate curve (i.e., the rate of deposition was not assumed to be constant between zone boundaries). These ages then were used to plot the pollen ratios and pinyon values in Figure 8. When Stuiver and Reimer (1986) became available, the conversions to tree-ring corrected calendar dates (Table II) were compared; the data comparison indicated that the zone 6 boundaries could be shifted slightly. However, based on the results of tree-ring correlation (which is discussed in the following sections), the original dates were retained.

Correlation and Relative Calibration of the La Plata Mountain Pollen Ratio Record with Tree-Ring Indices. In this study, long tree-ring records from high-elevation trees near timberline were sought as a climate proxy to be compared with that of the La Plata Mountain pollen records. A model of the relationship between tree growth and climate has been devised by Fritts (1976) that shows that stored food reserves are the link between the previous year's climate and current year's growth. That is, the food reserves (and its manifestation in a resultant tree-ring width) are constrained in trees growing at low elevations by moisture, whereas at high elevation, reserves are constrained primarily by summer temperature. In the latter case, the summer must be long enough or warm enough to allow production of adequate food reserves to be used for respiration and needle replacement requirements. In addition, the reserves must be sufficient for ring production the following year; the larger surplus values correlate to wider rings, so the warmer summers lead to wider rings.

LaMarche and Stockton (1974) obtained ring width records for high-elevation bristlecone pine (*Pinus aristata*) from Almagre Mountain (near Pikes Peak, Colorado). The Almagre record was selected for this study because it was the longest regional record that could be used to reconstruct summer warmth. The published yearly indices (Drew, 1974) were used to obtain the average for each successive 20-yr mean, which was plotted in Figure 8 on the 11th year. Changes in these tree-ring indices reflect long-term changes in summer temperature. A decrease in the indices indicates a lower relative summer temperature (or, for the purposes of this report, probably a shorter growing season; see Petersen (1988a) for a more complete discussion). Points of correlation between the Almagre and Twin Lakes records are shown by arrows in Figure 8.

In addition to the record from Almagre Mountain, many high-elevation bristlecone pine records for the western United States show very narrow rings for the mid-1800s (LaMarche and Stockton, 1974), coinciding with an intense cold period documented by deposition of Little Ice Age glacial moraine in the mountains of the western United States (Porter, 1986). Tree rings of timberline spruce trees in the La Plata Mountains also show very narrow rings for the mid-1800s (Petersen, 1988a). Scuderi (1990) reports narrow tree rings recorded for high-elevation fox tail pine (*Pinus balfouriana*) in the Sierra Nevadas of California for the mid-1800s, which he correlates to similarly aged lichen-dated glacial moraines.

As evident from Figure 8, the spruce forest in the La Platas was relatively narrow in the mid-1800s. According to LaMarche (1974), the tree-ring widths from bristlecone pines (in the White Mountains of eastern California), located near their upper and lower elevational ranges, are indicative of conditions that were relatively cold (based on narrow rings in high-elevation trees) and dry (based on narrow rings in low-elevation trees), compared to current conditions.

As discussed previously, pollen zone 6 begins at Twin Lakes with a very low ratio of conifer/NAP in A.D. 500s. Additional corroborative evidence for a widely spread cold episode during the 6th century is provided by Scuderi (1990), who reports narrow tree rings at that time in high-elevation fox tail pine records (for the Sierra Nevadas), which he correlates with similarly aged lichen-dated glacial moraines.

An Index of Pinyon, a Proxy for the Summer Monsoon. The bottom panel of Figure 8 plots the number of pinyon pollen grains falling on a 1 cm^2 surface area during 1 yr at Beef Pasture. These figures were calculated using the deposition rate curve based on Figure 7 to estimate the rate (cm/yr) applicable for each sample. This was obtained by calculating the slope of the tangent at the point where the sample depth intersects the deposition curve. This figure (cm/yr) was then multiplied by the pinyon pollen content from each sample (grains/cm^3) as estimated by using the *Lycopodium* spore tracers introduced for that purpose. This gave a yearly pinyon pollen influx value of pollen grains per square centimeter per year (grains/cm^2/yr).

Changes in pinyon pollen absolute influx values are evident for the past 200 yr; Figure 8 indicates an increase followed by a decrease. This reflects the actual history of pinyon trees on the landscape. Beginning in the last century, pinyon has been expanding its range in the Four Corners region (Erdman, 1970, p. 21; Spencer, 1964, p. 148; Van Pelt, 1978). That trend probably would have continued in the Great Sage Plain, except for the extensive clearing for farming. A 20-fold increase (from 105 to 1,988 grains/cm^2/yr) in pinyon pollen accumulation is evident at Beef Pasture for samples dated between A.D. 1750 and 1890. After 1890, pinyon pollen influx decreased steadily at Beef Pasture to 458 grains/cm^2/yr (a 4-fold decrease) in a sample dating to 1970. The record at Sagehen Marsh has even tighter chronological control and clearly records the same pattern (Petersen, 1985; Petersen *et al.*, 1987).

Summer rainfall seems critical for pinyon seedling establishment and tree growth. Physiological adaptations such as root characteristics, large seed, needle form and number, cuticle thickness, and small growth form all seem to be special adaptations that take advantage of summer precipitation (Daubenmire, 1943, p. 11; Emerson, 1932; Wells, 1979, p. 318). In addition, there is a coincidence between the geographical distribution of summer rainfall in the west and that of pinyon. Arguments are presented elsewhere (Petersen, 1985; Petersen, 1988a) to substantiate pinyon as a good index for long-term changes in the relative strength of the summer monsoon and Davis (this volume) provides some corroborative evidence for that conclusion. When the monsoon is weak, it arrives later, does not penetrate as far to the northwest (Figure 4), and leaves earlier. When it is strong, just the opposite is true. Figure 8 indicates that during the last 2,800 yr there have been episodes of the monsoon being weaker than the present, and between A.D. 750 and 1150 it was at least as strong as the historic period.

Correlation with Other Records. Schoenwetter (1966) was the first to propose a chronology of fluctuating winter-dominant precipitation during Anasazi occupation of the Four Corner region. He also suggested that the times of low winter precipiation probably were offset by increases of summer precipitation. His independently dated climatic sequence for effective moisture (Schoenwetter, 1966, 1967, 1970; Schoenwetter and Eddy, 1964), and those of Euler *et al.* (1979) and Dean *et al.* (1985) closely match the timing and direction of fluctuations presented here. (See Samuels and Betancourt, (1982) for an alternative explanation for the decrease of pinyon after A.D. 1150 in the Chaco Canyon region of northwest New Mexico.)

Another independently dated test of the Beef Pasture reconstruction is provided by the tallying of the pith dates of ponderosa pine (*Pinus ponderosa*) timbers that were used in construction by the Anasazi in the Dolores Project area. Dates of ponderosa pine seedlings establishment coincide with springs characterized by higher-than-average precipitation, whereas low spring precipitation or snowpack inhibit establishment (Schubert, 1974). Pith dates of archaeological tree-ring samples may not be able to be applied to the year of tree germination because there is no way of telling whether the sample represents the base of the trunk or a much higher (and therefore later) position on the bole. However, I attempted to overcome this by comparing 25-yr groupings of pith dates with the pollen record. (See Petersen, 1988a, for further details.) The cluster of Dolores pith dates before A.D. 900 (when the record ends) matches the periods reconstructed to be at least as wet at the present (a ratio value greater than 0.60 in Figure 8) (Petersen, 1988a). Additionally, the direction of vegetation change indicated by paired pollen samples from well-dated superimposed archaeological floors of differing age in the Dolores Project (Petersen, 1986) is in agreement with that proposed here.

Bird's Eye View of the Changing Width of the Potential Dry-Farming Belt. The Dolores Project Area is situated near the upper modern limits of the dry-farming belt; this belt, defined by both moisture and temperature, exists within narrow altitudinal limits (Figure 9a). Consequently, prehistoric agricultural activity should have been affected by changes in the elevational extent (width) of the agricultural belt as modern dry farming is today. The reconstructions presented in Figures 9 and 10 are based on an interpretation of the climate indices in Figure 8. The combination of Colorado Front Range tree-ring width data with the Twin Lakes conifers/NAP

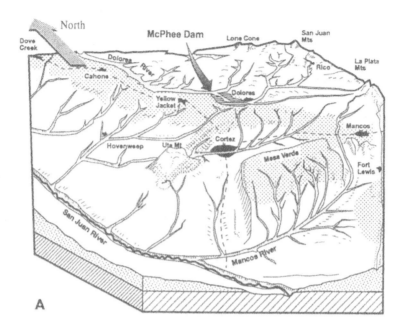

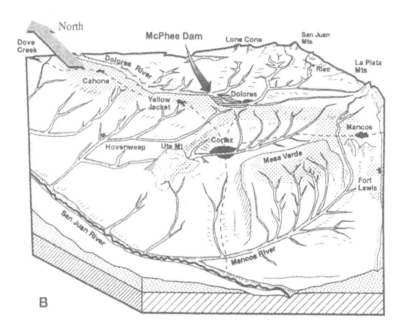

Fig. 9. A bird's eye view of the reconstruction of the relative width of the potential dry-farm beld in southwestern Colorado. (A) Modern and A.D. 600–800, (B) A.D. 800–1000 and 1100–1300.

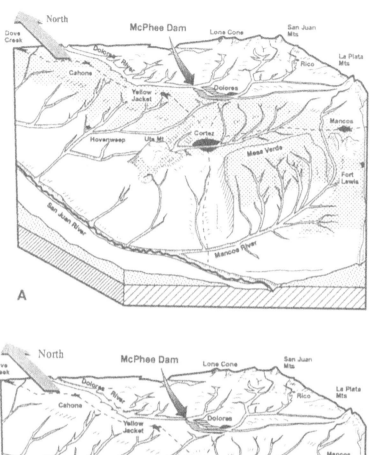

Fig. 10.   A bird's eye view of the reconstruction of the relative width of the potential dry-farm belt in southwestern Colorado. (A) A.D. 1000–1100, (B) A.D. 1300–1850.

pollen ratio is used to locate the upper elevational limit of the dry-farm belt and described in terms of growing season length. The Beef Pasture spruce/pine ratios are used to locate the lower elevational extent of the dry-farming belt (supplemented with archaeological site distribution data as discussed below) and described in terms of jet-stream derived winter precipitation. Finally, pinyon pollen influx is used to evaluate the risk for dry farming within the farm belt and also is used to characterize the strength of the summer monsoon. The pattern shown in Figure 9a is the historic pattern and represents the reconstruction for the period A.D. 600–800.

At the beginning of the A.D. 600–800 period, after the very narrow spruce zone centered about A.D. 550, the Almagre tree-rings show a very rapid increase in width. The region was emerging from the grips of cold weather of the 6th century. At that time the Anasazi began appearing in large numbers in southwest Colorado (Schlanger, 1988). The rapid warming widened the dry-farming belt to about the same size it is today. The pollen samples dating to the mid A.D. 600s from Sagehen Marsh (Table II) in the Dolores Project area are indistinguishable from the pollen samples that date shortly before the 1920 clearing (Petersen, 1984, 1988b; Petersen *et al.*, 1987). During this period populations grew and the Anasazi thrived.

About A.D. 750, long-term winter drought moved into the area and the agricultural zone narrowed by about half of that today (Figure 9b). That winter drought continued for about 200 yr; however, it was offset somewhat by increased summer convective storms that more commonly discharged precious moisture near the elevation of the potential dry-farm belt, rather than at lower elevations (Farmer and Fletcher, 1971; Petersen *et al.*, 1987). The large population increases that occurred in the Dolores Project area at that time cannot be accounted for by births alone, suggesting that the Anasazi survived by moving their fields to higher elevations and exploiting the narrowed (but still productive) farm-belt (Schlanger, 1988). Historic documents show that about half the farm land in the region (and those from mostly lower elevations in the historic farm belt) was abandoned during the severe drought of the 1930s, providing an analog for a narrowed dry-farm belt (Gregory and Thorpe, 1938; Petersen *et al.*, 1987).

There is a period of narrow tree-ring widths in the Almagre record that dates to the 10th century. A further test of the correlation between narrow tree-ring and summer warmth or growing season length is provided by Dolores Project archaeological data. Areas subject to cold air pooling were mapped in the Dolores Project area, and it was found that although these areas had been almost continuously occupied since the 7th century, they were abandoned by Anasazi farmers for a short time in the 10th century when the Almagre tree-ring record suggests cooling. The farm land was again reoccupied in the 11th century when tree-ring width at timberline again became wider (Breternitz *et al.*, 1986; Petersen *et al.*, 1987).

By A.D. 1000, the climate had warmed to a point that the growing season was adequate for farming in the Dolores Project area, and the summer monsoon moisture was supplemented by a large increase in the amount of winter precipitation. This combination greatly expanded the farming belt, dropping its lower limit to about

1,600 m elevation (Figure 10a), which more than doubled the amount of farmland available, compared to that of today (as shown in Figure 9a). During this time period, there was a tremendous explosion in archeological evidence (locally called the PII expansion) for areas currently below the modern extent of the farm belt.

It is generally accepted that Anasazi dwellings would not be located too far from their corn fields. For instance at Hovenweep National Monument (1,600 m), where there are abundant Anasazi ruins, one cannot grow corn today because the 30 cm of annual precipitation at that elevation is not adequate. In fact, modern farmers cannot obtain crop insurance for dry farming for elevations below 1,830 m (some 230 m higher in elevation) and routine dry farming is not practice below 2,010 m (Petersen *et al.*, 1987). However, Woosley (1977) found in surface pollen transect on suspected Anasazi dry-farming fields at Hovenweep that there was corn and bean pollen still preserved in the soil by the arid climate. Corn pollen is a relatively large pollen grain that usually does not travel more than a few meters from the source plant and even in modern corn fields is only 1 or 2% of the pollen count (Martin, 1963, p. 50). To find it in such upland field locations suggests that conditions for rainfall farming had to have been drastically different from current conditions. Possibly the combination of the high winter snowpack like the early part of pollen zone 6 (Figure 8) but with added component of additional summer precipitation combined to produce the unprecedented broad farm belt. It was as if there was greater vigor in the general circulation regime whereby more oceanic climate (both summer and winter) were pulled deeper inland (see Wallen (1955) for a possible explanation for such an occurrence earlier this century).

During the period from A.D. 1000 to A.D. 1100 the Anasazi cultures reached their zenith. Populations incrased dramatically in the Four Corners region and else-where (Figure 1) and developed complex social, economic, and political structures, the most famous of which is preserved in Chaco Canyon National Monument in northeastern New Mexico. The monsoon boundary (Figure 2) most likely was located north of its present location.

Interestingly the expansion of the farm belt at its lower elevational limit in the Four Corners region actually led to a decrease in population in the Dolores Project area (Schlanger, 1988), now located at the uppermost margin. This was most likely because the Anasazi farmers now had a greater choice of land and could avoid the cold air drainage that often occurs in the Dolores River valley proper (Petersen *et al.*, 1987).

Soon after A.D. 1100, another winter drought began to move into the region, narrowing the potential dry-farm belt from the bottom (Figure 9b). As before, the Anasazi adjusted, but unlike the drought of the A.D. 800s (Figure 8), summer rain was not as plentiful or dependable, the summer monsoon boundary was most likely located south of its present locations. Summer rains arrived later, left earlier, and were not as predictable. This time the Anasazi compensated by utilizing a number of water control strategies such as check dams, ditches, and reservoirs (e.g., Erdman *et al.*, 1969).

Then severe, dry cold started to move into the region about A.D. 1200, making farming risky at higher elevations (Figure 8). As the summer and winter drought pinched the farming belt from the bottom, cold pinched it from the top. By A.D. 1200 the Anasazi began leaving the area. The most severe impact of the combination was felt late in the A.D. 1200s, where in effect, the farming belt was squeezed to the point that it disappeared (Figure 10b). The Anasazi simply left the higher elevations of the Four Corners region and headed south, seeking more dependable summer monsoons, sufficient winter precipitation (or areas that could be irrigated), and longer growing seasons (Figure 1).

Interestingly, the cold, dry conditions that began in the A.D. 1200s lasted for hundreds of years – without much change – to about the A.D. 1850s. Because weather stations in the region were mostly established after 1895, somewhat after the change to warmer and wetter conditions that have existed during the last 130 years or so, our modern weather records give little hint of the severity of conditions that had occurred. The narrow tree rings in timberline trees in the La Plata Mountains during this time period is suggestive (Petersen, 1988a). Further, an account by a military expedition led by Col. Macomb into the region in 1859 describes relatively cold conditions. Newberry (1876, pp. 76–77, pp. 88–89) observed a broad surface of snow covering the San Miguel Mountains of the San Juans in early August (today it is gone by early July), and from a site near modern day Yellow Jacket they observed that it was *too high, too cold,* and *too dry* to grow corn and other crops in the area. This same locality today supports thousands of hectares of successful dry farming of those very same crops (Petersen *et al.,* 1987; Petersen, 1988a).

In the 1870s, some 20 to 30 yr after Macomb's and Newberry's expedition, when white settlers began moving into the area (Figure 9a), the mountain peaks were clear of snow in late summer, young pinyon trees were invading the sagebrush, the spruce forest was beginning to expand, and timberline advance to higher elevations and the rings in timberline trees became wider. All these were signs that the farming belt had rebounded and that the new settlers would find ideal dry-farming conditions with abundant summer and winter precipitation and an adequate growing season. The climate had reverted to conditions similar to those during the 700-yr Anasazi occupation, and the modern farms could locate exactly where the Anasazi had farms during their sojourn in the region.

5. Conclusions

In the next century, increases in atmospheric trace gas concentration could warm the global average temperature beyond what it has ranged during the past century. Examination of larger-than-historic climatic changes that have occurred in the past in specific regions provides realistic context for evaluating such potential future changes. This paper has contrasted the climatic manifestation of the Little Climatic Optimum or Medieval Warm Period (A.D. 900–1300) with that of the Little Ice

Age (A.D. 1300–1850) in the northern Colorado Plateau region of the southwestern U.S.A. The zenith of the Anasazi occupation coincides with the former and their demise coincides with the latter, when conditions became too cold and especially dry (in the summer) to support upland dry farming. During the height of the Little Climatic Optimum the region was characterized by a relatively long growing season and greater winter and summer precipitation than that of today. This resulted in a relatively rapid development of a potential dry-farming belt that was twice as wide as the present and areas that cannot be dry farmed today were routinely farmed by the Anasazi. Such conditions would be beneficial to dry farmers in the Four Corners region if those conditions were repeated in the near future.

Acknowledgements

Work on climatic change in the Colorado Rockies grew out of an interest fostered while I was a field assistant to James B. Benedict in the Colorado Front Range in the mid 1960s. Subsequently I studied a number of fruitful years with Peter J. Mehringer, first at the University of Utah, and later at Washington State University, and I studied one rewarding year with Paul S. Martin at the University of Arizona. The work reported here began in 1972 in conjunction with the Salmon Ruins Archaeological Project (supported in part by National Endowments for the Humanities grants RO-8637A-74-540 and RO-24580-76-630 to Cynthia Irwin-Williams) and continued during the execution of the Dolores Projects's Dolores Archaeological Program funded by the Bureau of Reclamation, Upper Colorado Region, U.S. Department of Interior (Contract No. 8-07040-S0562, David A. Breternitz senior principal investigator). Additional work has been undertaken during my employment by the Westinghouse Hanford Company, Richland, Washington (Hanford Operations and Engineering Contractor for the U.S. Department of Energy under Contract DE-AC06-87RL10930) in programs for developing defensible estimates of potential future climate for the Hanford Site and the DOE's Monticello (Utah) Remedial Action Project for uranium mill tailings. I thank J. B. Benedict and O. K. Davis for throughtful reviews. The views and opinions of authors expressed herein do not necessarily state or reflect those of the United States Government or any agency thereof and the article is not subject to copyright.

References

Barry, R. G. and Bradley, R. S.: 1976, 'Historical Climatology', in Steinhoff, H. W. and Ives, J. D. (eds.), *Ecological Impacts of Snowpack Augmentation in the San Juan Mountains, Colorado*, San Juan Ecology Project, Final Report, Colorado State University Publications, Fort Collins, pp. 43–47.
Bray, J. R.: 1971, 'Vegetationed Distribution, Tree Growth and Crop Success in Relation to Recent Climatic Change', in Cragg, J. B. (ed.), *Advances in Ecological Research* **7**, pp. 177–233.
Betancourt, J. L.: 1984, 'Late Quaternary Plant Zonation and Climate in Southeastern Utah', *Great Basin Natural.* **44**, 1–35.

Berlin, G. L., Ambler, J. R., Hevly, R. H., and Shaber, G. G.: 1977, 'Identification of a Sinagua Agricultural Field by Aerial Thermography, Soil Chemistry, Pollen/Plant Analysis, and Archaeology', *Amer. Antiq.* **42**, 588–600.
Bradley, R. S. and Barry, R. G.: 1973, 'Secular Climatic Fluctuations in Southwestern Colorado', *Month. Wea. Rev.* **101**, 264–270.
Breternitz, D. A., Robinson, C. K., and Gross, G. T., (compilers): 1986, *Dolores Archaeological Program: Final Synthetic Report*, U.S. Department of Interior, Bureau of Reclamation, Engineering and Research Center, Denver.
Brown, D. E. (ed.): 1982, 'Biotic Communities of the American Southwest – United States and Mexico', *Desert Plants* **4**(1–4), 1–342.
Bryson, R. A. and Julian, P. R. (conveners): 1963, *Proceedings of the Conference on the Climate of the Eleventh and Sixteenth Centuries (Aspen, Colorado, June 16–24, 1962)*, NCAR Technical Notes 63–1, AFCRL–63–660, National Center for Atmospheric Research, Boulder, Colorado.
Bryson, R. A. and Lowry, W. P.: 1955, 'Synoptic Climatology of the Arizona Summer Precipitation Singularity', *Bull. Amer. Meteorol. Soc.* **36**, 329–339.
Cordell, L. S.: 1984, *Prehistory of the Southwest*, Academic Press, Orlando, Florida.
Damon, P. E., Ferguson, C. W., Long, A., and Wallick, E. I.: 1974, 'Dendrochronologic Calibration of the Radiocarbon Time Scale', *Amer. Antiq.* **39** (2 pt.1), 350–366.
Daubenmire, R. F.: 1943, 'Soil Temperature versus Drought as a Factor Determining Lower Altitudinal Limits of Trees in the Rocky Mountain', *Botan. Gaz.* **105**, 1–13.
Daubenmire, R. F.: 1954, 'Alpine Timberlines in the Americas and Their Interpretation', *Butler Univ. Botan. Stud.* **11**, 119–136.
Dean, J.S., Euler, R. C., Gumerman, G. J., Plog, F., Hevly, R. H., and Karlstom, T. H. V.: 1985, 'Human Behavior, Demography, and the Paeoenvironment of the Colorado Plateaus', *Amer. Antiq.* **50**, 537–554.
Dix, R. L. and Richards, J. L.: 1976, 'Possible Changes in Spectra Structure of the Subalpine Forest Induces by Increased Snow Pack', in Steinhoff, H. W. and Ives, J. D. (eds.), *Ecological Impacts of Snowpack Augmentation in the San Juan Mountains, Colorado*, San Juan Ecology Project, Final Report, Colorado State University Publications, Fort Collins, pp. 311–322.
Dorrah, J. H., Jr.: 1946, *Certain Hydrological and Climatic Characteristics of the Southwest*, Publications in Engineering No. 1, University of New Mexico, Albuquerque.
Drew, L. G. (ed.): 1974, *Tree-Ring Chronologies of Western America: IV, Colorado, Utah, Nebraska, and South Dakota*, Chronology Series 1, University of Arizona Laboratory of Tree-Ring Research, Tucson.
Emerson, F. W.: 1932, 'The Tension Zone Between the Grama Grass and Pinyon-Juniper Associations in Northeastern New Mexico', *Ecology* **13**, 347–358.
Erdman, J. A.: 1970, 'Pinyon-Juniper Succession after Natural Fires on Residual Soils of Mesa Verde, Colorado', *Brigham Young Univ. Sci. Bull., Biol. Ser.* **11**, 17–26.
Erdman, J. A., Douglas, C. L., and Marr, J. W.: 1969, *Wetherill Mesa Studies: Environment of Mesa Verde, Colorado*, Archaeological Research No. 7B, U.S. Department of Interior, National Park Service, Washington, D.C.
Euler, R. C., Gumerman, G. J., Karlstrom, T. N. V., Dean, J. S., and Hevly, H. R.: 1979, 'The Colorado Plateaus: Cultural Dynamics and Paleoenvironment', *Science* **205**, 1089–1101.
Farmer, E. E. and Fletcher, J. E.: 1971, *Precipitation Characteristics of Summer Storms at High-Elevation Stations in Utah*, Forest Service Research Paper INT-110, U.S. Department of Agriculture, Forest Service, Ogden.
Fritts, H. C.: 1976, *Tree Rings and Climate*, Academic Press, New York.
Fritts, H. C., Smith, D. G., and Stokes, M. A.; 1965, 'The Biological Model for Paleoclimatic Interpretation of Mesa Verde Tree-Ring Series', in Osborne, D. (assembler), *Contributions of the Wetherill Mesa Archaeological Project*, Society of American Archaeology Memoirs 19, Salt Lake City, pp. 101–121.
Gregory, H. E. and Thorpe, M. R.: 1938, *The San Juan Country: A Geographic and Geologic Reconnaissance of Southeastern Utah*, Professional Paper 188, U.S. Geological Survey, U.S. Government Printing Office, Washington, D.C.
Grove, J. M.: 1988, *The Little Ice Age*, Methuen, London.

Gumerman, G. J. (ed.): 1988, *The Anasazi in a Changing Environment*, Cambridge University Press, New York.

Heermann, D. F., Finkner, M. D., and Hiler, E. A.: 1971, *Probability of Sequences of Wet and Dry Days for Eleven Western States and Texas*, Experiment Station Technical Bulletin 117, Colorado State University, Fort Collins.

Hevly, R. H.: 1981, 'Pollen Production, Transport and Preservation: Potentials and Limitations in Archaeological Palynology', *J. Ethnobiol.* **1**, 39–54.

Houghton, J. T., Jenkins, G. J., and Ephraums, J. J. (eds.): 1990, *Climate Change: The IPCC Scientific Assessment*, Cambridge University Press, Cambridge, U.K.

Huntington, E.: 1914, *The Climatic Factor as Illustrated in Arid America*, Publication 192, Carnegie Institution of Washington, Washington, D.C.

Jennings, J. D.: 1968, *Prehistory of North America*, McGraw-Hill, New York.

King, J. E.: 1967, 'Modern Pollen Rain and Fossil Pollen in Soils in the Sandia Mountains, New Mexico', *Papers Michig. Acad. Sci., Arts Lett.* **52**, 31–41.

Klein, J., Ierman, J. C., Damon, P. E., and Ralph, E. K.: 1982, 'Calibration of Radiocarbon Dates: Tables Based on the Consensus Data of the Workshop on Calibrating the Radiocarbon Time Scale', *Radiocarbon* **24**, 103–150.

LaMarche, V. C., Jr.: 1974, 'Paleoclimatic Inferences from Long Tree-Ring Records', *Science* **183**, 1043–1048.

LaMarche, V. C., Jr., and Stockton, C. W.: 1974, 'Chronologies from Temperature Sensitive Bristlecone Pines at Upper Treeline in the Western United States', *Tree-Ring Bull.* **34**, 21–45.

Lamb, H. H.: 1977, *Climate: Present, Past and Future, Vol. 2, Climate History and the Future*, Methuen, London, England.

Maher, L. J., Jr.: 1963, 'Pollen Analyses of Surface Materials from the Southern San Juan Mountains, Colorado', *Geol. Soc. Amer. Bull.* **74**, 1485–1504.

Malaurie, J., Vasari, Y., Hyvarinen, H., Delibrias, G., and Labeyrie, J.: 1972, 'Preliminary Remarks on Holocene Paleoclimates in the Regions of Thule and Inglefield Land, above A11 since the Beginning of Our Own Era', in Vasari, Y., Hyvarinen, H., and Hicks, S. (eds.), *Climate Changes in Arctic Areas during the Last Ten-Thousand Years: A Symposium Held at Oulanka and Kevo, 4–10 October 1971*, Acta Universitatis Ouluensis, Series A, Scientiae Refum Naturalium No. 3, Geologica No. 1, University of Oulu, Oulu, Finland, pp. 105–133.

Martin, P. S.: 1963, *The Last 10,000 Years: A Fossil Pollen Record of the American Southwest*, The University of Arizona Press, Tucson.

Martin, P. S., and Byers, W.: 1965, 'Pollen and Archaeology at Wetherill Mesa', in Osborne, D. (assembler), *Contributions of the Wetherill Mesa Archaeological Project*, Society of American Archaeology Memoirs 19, Salt Lake City, pp. 101–121.

Miller, J. F., Frederick, R. H., and Tracey, R. J.; 1973, *Precipitation – Frequency Atlas of the Western United States, Volume III – Colorado: NOAA Atlas 2*, U.S. Department of Commerce, National Oceanic and Atmospheric Administration, National Weather Service, Silver Springs, Maryland.

Mitchell, V. L.: 1976, 'Regionalization of Climate in the Western United States', *J. Appl. Meteorol.* **15**, 920–927.

Newberry, J. S.: 1876, 'Geological Report', in Macomb, J. N. (ed.), *Report of the Exploring Expedition from Santa Fe, New Mexico, to the Junction of the Grand and Green Rivers of the great Colorado of the West, in 1859, under the Command of Capt. J. N. Macomb, Corps of Topographical Engineers (now Colonel of Engineers); with Geological Report by Prof. J. S. newberry, Geologist of the Expedition*, Engineer Department, U.S. Army, Government Printing Office, Washington, D.C.

Nichols, H.: 1975, *Palynological and Paleoclimatic Study of the Late Quaternary Displacement of the Boreal Forest-Tundra Ecotone in Keewatin and Mackenzi, N.W.T., Canada*, Occasional Paper 15, Institute of Arctic and Alpine Research, Boulder, Colorado.

Osborne, D. (assembler): 1965, *Contributions of the Wetherill Mesa Archaeological Project*, Society of American Archaeology Memoirs 19, Salt Lake City.

Patzelt, G.: 1974, 'Holocene Variations of Glaciers in the Alps, in *les Methodes Quantitative d'Etude des Variations du Climat au cours du Pleistocene, Gif-sur-Yvette*, 5–9 juin 1973, Colloques Internationaux du Centre National de la Recherche Scientifique **219**, 51–59.

Pearson, G. A.: 1931, *Forest Types in the Southwest as Determined by Climate and Soil*, Technical Bulletin 247, U.S. Department of Agriculture, U.S. Government Printing Office, Washington, D.C.

Van Pelt, N. S.: 1978, 'Woodland and Parks in Southeastern Utah', unpublished Masters' Thesis, Department of Geography, University of Utah, Salt Lake City.

Petersen, K. L.: 1984, 'Man and Environment in the Dolores River Valley, S. W. Colorado: Some Pollen Evidence', in *AMQUA 1984, Program and Abstracts*, Eighth Biennial Meeting, American Quaternary Association, University of Colorado, Boulder, p. 102.

Petersen, K. L.: 1985, 'Palynology in Montezuma County, Southwestern Colorado: The Local History of Pinyon Pine (*Pinus edulis*)', in Jacobs, B., Davis, O., and Fall, P. (eds.), *Late Quaternary Palynology of the American Southwest*, ASSP Contribution Series 16, American Association of Stratigraphic Palynologists Foundation, Dallas, pp. 47–62.

Petersen, K. L.: 1986, 'Section 3: Pollen Studies: Temporal Patterns and Resource Uses', in Petersen, K. L., Matthews, and Neusius, S. W., 'Chapter 4 – Environmental Archaeology', in Breternitz, D. A., Robinson, C. K., and Gross, G. T. (compilers), *Dolores Archaeological Program: Final Synthetic Report*, U.S. Department of Interior, Bureau of Reclamation, Engineering and Research Center, Denver, pp. 184–199.

Petersen, K. L.: 1988a, *Climate and the Dolores River Anasazi*, University of Utah Anthrolopological Papers No. 113, University of Utah Press, Salt Lake City.

Petersen, K. L.: 1988b, 'Comparison of Modern Surface Pollen Samples with Samples from Sagehen Marsh, Dolores River Valley, Montezuma County, Southwestern Colorado', *AMQUA 1988, Program and Abstracts*, Tenth Biennial Meeting, American Quaternary Association, University of Massachusetts, Amherst, p. 147.

Petersen, K. L. and Mehringer, P. J., Jr.: 1976, 'Postglacial Timberline Fluctuations, La Plata Mountains, Southwestern Colorado', *Arctic Alpine Res.* **8**, 275–288.

Petersen, K. L., Clay, V. L., Matthews, M. H., Neusius, S. W. (compilers), and Breternitz, D. A. (principal investigator): 1985, *Dolores Archeological Program: Studies in Environmental Archaeology*, U.S. Department of Interior, Bureau of Reclamation, Engineering and Research Center, Denver.

Petersen, K. L., Orcutt, J. D. (compilers), and Breternitz, D. A. (principal investigator): 1987, *Dolores Archaeological Program: Supporting Studies: Settlement and Environment*, U.S. Department of Interior, Bureau of Reclamation, Engineering and Research Center, Denver.

Porter, S. C.: 1986, 'Pattern and Forcing of Northern Hemisphere Glacier Variations during the Last Millennium', *Quatern. Res.* **26**, 27–48.

Pyke, C. B.: 1972, *Some Meteorological Aspects of the Seasonal Distribution of Precipitation in the Western United States and Baja California*, Water Resources Center Contribution No. 139, University of California, Los Angeles.

Samuels, M. L. and Betancourt, J. L.: 1982, 'Modeling the Long-Term Effects of Fuelwood Harvest on Pinyon-Juniper Woodlands', *Environm. Managem.* **6**, 505–515.

Schlanger, S. H.: 1988, 'Patterns of Population Movement and Long Term Population Growth in Southwestern Colorado', *Amer. Antiq.* **35**, 773–793.

Schneider, S. H.: 1986, 'Can Modeling of the Ancient Past Verify Prediction of Future Climates? An Editorial', *Clim. Change* **8**, 117–119.

Schoenwetter, J.: 1966, 'A Re-Evaluation of the Navajo Reservoir Pollen Chronology', *El Palacio* **73**, 19–26.

Schoenwetter, J.: 1967, 'Pollen Survey of the Shiprock Area', in Harris, A. H., Schoenwetter, J., and Warren, A. H., *An Archaeological Survey of the Chuska Valley Chaco Plateau, New Mexico. Part I: Natural Science Studies*, Research Records 4, Museum of New Mexico, Albuquerque, pp. 72–103.

Schoenwetter, J.: 1970, 'Archaeological Pollen Studies of the Colorado Plateau', *Amer. Antiq.* **35**, 35–48.

Schoenwetter, J. and Eddy, F. W.: 1964, *Alluvial and Palynological Reconstruction of Environments, Navajo Reservoir District*, Papers in Anthropology 13, Museum of New Mexico, Albuquerque.

Schubert, G. H.: 1974, *Silviculture of Southwestern Ponderosa Pine: The Status of Our Knowledge*, Forest Service Research Paper RM-123, U.S. Department of Agriculture, Forest Service, Fort Collins.

Scuderi, L. A.: 1990, 'Tree-Ring Evidence for Climatically Effective Volcanic Eruptions', *Quatern. Res.* **34**, 67–85.

Short, S. K. and Nichols, H.: 1977, 'Holocene Pollen Diagrams from Subarctic Labrador-Ungava: Vegetational History and Climate Change', *Arctic Alpine Res.* **9**, 265–290.

Smiley, T. L.: 1961, 'Evidences of Climatic Fluctuations in Southwestern Prehistory', in Fairbridge, R. W. (ed.), *Solar Variations, Climatic Change, and Related Geophysical Problems*, Annals of the New York Academy of Sciences 95(Art. 1), New York, pp. 697–704.

Spencer, D. A.: 1964, 'Porcupine Population Fluctuations in Past Centuries Revealed by Dendrochronology', *J. Appl. Ecol.* **1**, 127–159.

Stockton, C. W.: 1976, 'Long-Term Streamflow Reconstruction in the Upper Colorado River Basin using Tree Rings', in Clyde, C. A., Falkenburg, P. H., and Riley, J. P. (eds.), *Colorado River Basin Modeling Studies: Proceedings of a Seminar Held at Utah State University, College of Engineering, Logan, Utah*, July 16–18, 1975, Utah Water Research Laboratory, Utah State University, pp. 401–441.

Stuiver, M. and Reimer, P. J.: 1986, 'A Computer Program for Radiocarbon Age Calibration', *Radiocarbon* **28**, 1022–1030.

Thomas, H. E.: 1959, 'Reservoirs to Match our Climatic Fluctuation', *Amer. Meteorol. Soc. Bull.* **40**, 240–249.

Tranquillini, W.: 1979, *Physiological Ecology of the Alpine Timberline: Tree Existence at High Altitudes with Special Reference to the European Alps*, Springer-Berlag, Berlin

U.S. Department of Agriculture, Soil Conservation Serive: 1976, *Land Use and Natural Plant Communities, Montezuma County, Colorado*, Map M7-0-23444-43, U.S. Department of Agriculture, Soil Conservation Service, Portland.

Wallen C. C.: 1955, 'Some Characteristics of Precipitation in Mexico', *Geograf. Annal.* **37**, 51–85.

Wardle, P.: '1974, 'Alpine Timberlines', in Ives, J. D. and Barry, R. G. (eds.), *Arctic and Alpine Environments*, Methuen, London, pp. 371–402.

Wells, P. V.: 1979, 'An Equable Glaciopluvial in the West: Pleniglacial Evidence of Increased Precipitation on the Gradient from the Great Basin to the Sonoran and Chihuahuan Deserts', *Quatern. Res.* **12**, 311–325.

Woodbury, R. W.: 1961, 'Climatic Changes and Prehistoric Agriculture in the Southwestern United States', in Fairbridge, R. W. (ed.), *Solar Variations, Climatic Change, and Related Geophysical Problems*, Annals of the New York Academy of Sciences 95(Art. 1), New York, pp. 697–704.

Woosley, A. I.: 1977, 'Farm Field Location through Palynology', in Winters, J. C. (ed.), *Hovenweep 1976*, Archaeology Report No. 3, Anthropology Department, San Jose State University, San Jose, California, pp. 133–150.

Wormington, H. M.: 1947, *Prehistoric Indians of the Southwest*, Popular Series 7, Denver Museum of Natural History, Denver.

Wright, J. W.: 1952, *Pollen Dispersion of Some Forest Trees*, Northeastern Forest Experiment Station Paper 46, U.S. Department of Agriculture, Forest Service, Upper Darby, Pennsylvania.

(Received 22 September, 1992; in revised form 19 October, 1993)

THE CORRELATION OF SUMMER PRECIPITATION IN THE SOUTHWESTERN U.S.A. WITH ISOTOPIC RECORDS OF SOLAR ACTIVITY DURING THE MEDIEVAL WARM PERIOD

OWEN K. DAVIS

Department of Geosciences, University of Arizona, Tucson, AZ 85721, U.S.A.

Abstract. Decreased solar activity correlates with positive cosmogenic isotope anomalies, and with cool, wet climate in temperate regions of the world. The relationship of isotope anomalies to climate may be the opposite for areas influenced by monsoonal precipitation, i.e., negative anomalies may be wet and warm. Petersen (1988) has found evidence for increased summer precipitation in the American Southwest that can be shown to be coincident with negative ^{14}C anomalies during the Medieval Warm Period. The present study compares palynological indicators of lake level for the Southwest with Petersen's data and with the ^{14}C isotope chronology. Percentages of aquatic pollen and algae from three sites within the Arizona Monsoon record greater lake depth or fresher water from A.D. 700–1350, between the Roman IV and Wolf positive isotope anomalies, thereby supporting Petersens's findings. Maximum summer moisture coincides with maximum population density of prehistoric people of the Southwest. However, water depth at a more northern site was low at this time, suggesting a climate–isotope relationship similar to that of other temperate regions. Further analysis of latitudinal patterns is hampered by inadequate ^{14}C dating.

1. Introduction

1.1. *Rapid Climate Change*

The potential for large climatic change ($> 1\,°C$, > 100 mm precipitation) in response to increased concentratrions of greenhouse gasses has focused the attention of the scientific community on rapid climatic change (decade – century). Studies of the potential effects primarily have been in two areas: computer simulations of global climate (GCM's) under increased concentrations of greenhouse gasses, and investigation of analogous periods of rapid climate change, particularly rapid warming (e.g. Smith and Tirpak, 1989). The GCM's have the advantage of targeting a causative variable (e.g., CO_2); but the results of inter-model comparisons show dramatic variability, particularly for moisture in mid-latitudes. The latter approach is less exact because the histories of atmospheric gasses are poorly known, and because concentrations of greenhouse gasses appear to covary with other climate-forcing phenomena such as incoming solar radiation (Barnola *et al.*, 1987). However, the analog approach has the distinct advantage of demonstrating how plants and animals *have* responded to rapid changes in temperature and precipitation.

Climatic Change **26**: 271–287, 1994.
© 1994 *Kluwer Academic Publishers.*

1.2. *Causes of the Medieval Warm Period*

The climatic factors that led to the Medieval Warm Period (**MWP**) are not fully known. Based on polar ice cores, the atmospheric CO_2 concentration during the MWP was no higher than the pre-industrial average (Post *et al.*, 1990, figure 3b). Therefore, the value of the MWP as an analog is as a record of the biotic response to warming rather than as an example of climatic response to greenhouse forcing. The number of volcanic eruptions was low from A.D. 1100–1600 (Bryson, 1988), so reduced atmospheric aerosols may have permitted increased solar radiation at the earth's surface. In addition, solar radiation at the top of the atmosphere may have been greater during the MWP.

The association between increased solar activity and the MWP has been made by Eddy (1977) and by Hood and Jirikowic (1990) who use the concentration of ^{14}C in tree rings as an index of solar activity. They refer to the interval of low ^{14}C production from A.D. 1120–1280 as the 'Medieval' or 'Grand' cosmogenic isotope anomaly. Likewise, Schmidt and Gruhle (1988) note the occurrence of drought in western Europe during low ^{14}C production periods of the Medieval Warm Period (11th to 13th centuries), the early Iron Age (200 B.C.–300 A.D.), and the Bronze Age (2000–1000 B.C.).

1.3. *Cosmogenic Isotopes, Solar Activity and Climate*

The production of cosmogenic isotopes (^{14}C and ^{10}Be) is controlled by the influence of the solar wind and the earth's magnetic field on the rate of cosmic ray bombardment of nitrogen and oxygen in the upper atmosphere (Figure 1). An inactive sun results in a weak solar wind, increased cosmic ray bombardment of Earth's upper atmosphere, and greater ^{14}C and ^{10}Be production; and its lower irradiance results in global cooling (Eddy, 1977). The climatic response may further influence ^{14}C record by modifying the carbon cycle (Jirikowic *et al.*, 1993). A strong geomagnetic field deflects cosmic rays and, therefore, reduces the rate of ^{14}C and ^{10}Be production. Geomagnetic modulation is responsible for the long-term trend in isotopes shown in Figure 2; and Stuiver *et al.* (1991) have suggested it also could be responsible for the shorter (2200 yr) variation. High-frequency variation (i.e., 208 and 88 yr cycles) arises from changes in solar activity (Damon and Sonett, 1991; Stuiver *et al.*, 1991).

The best records of cosmogenic isotope variation come from the ^{14}C dating of tree rings (Stuiver and Pearson, 1986; Pearson and Stuiver, 1986; Pearson *et al.*, 1986; Linick *et al.*, 1985; Stuiver *et al.*, 1986; Kromer *et al.*, 1986; Linick *et al.*, 1986). The timing and magnitude of the high frequency $\Delta^{14}C$ fluctuations match the record of cosmogenic ^{10}Be from polar ice cores (Beer *et al.*, 1990).

The connection between cosmogenic isotope anomalies and climate first was postulated by de Vries (1958) and was further developed by Suess (1965), Damon (1968), Eddy (1977), Stuiver (1965), and Raisbeck *et al.* (1990). Suess (1965) and

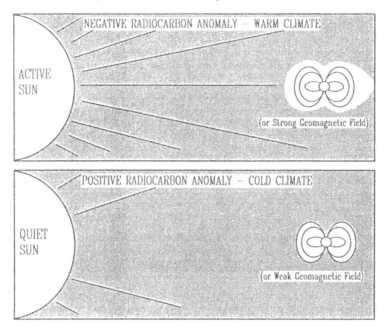

Fig. 1. Relationship of solar activity and the Earth's geomagnetic field to ^{14}C and [10]Be production.

de Vries (1958) noted the correspondence between reduced ^{14}C production, colder global climate, and reduced solar activity during the Maunder (sunspot) minimum, A.D. 1645–1715, when several lines of evidence indicate that solar activity was reduced by about 0.1% (Eddy, 1976; Lean, 1991), and global temperature was ca. 1 °C less than today (Grove, 1988). During the Maunder and other Little Ice Age sunspot minima (the Dalton A.D. 1400–1510, and Wolf A.D. 1290–1350), the radiocarbon content of the atmosphere was ca. 10‰ greater than background values (e.g., Pearson *et al.*, 1986).

The positive ^{14}C anomalies are named 'minima' by Eddy (1977) due to the correlation of increased ^{14}C production during the Little Ice Age with *minimum* sunspot numbers; e.g., the Maunder sunspot minimum. Conversely, the intervals between the positive anomalies are called 'maxima' by Eddy (1977), who correlates them with warm periods; for example the 'Medieval Maximum (1120–1280 A.D.)' and the MWP. The maxima in the ^{14}C record are less pronounced than the minima, but both are associated with brief climatic events (Figure 2).

The association of cold, wet periods and ^{14}C minima has been made by several authors (Eddy, 1977; Davis, 1992; Davis *et al.*, 1992; Jirikowic *et al.*, 1993; Schmidt and Gruhle, 1988; Davis and Shafer, 1992) and is most apparent for large anomalies in temperate regions, for example, the Homeric and Greek anomalies (Figure 2). Smaller minima, such as those near the MWP, have fewer associated climatic events, although the Wolf (745–620 yr), and Medieval (1020–880 yr) minima coincide with cool, wet periods in southern California (Davis *et al.*, 1992;

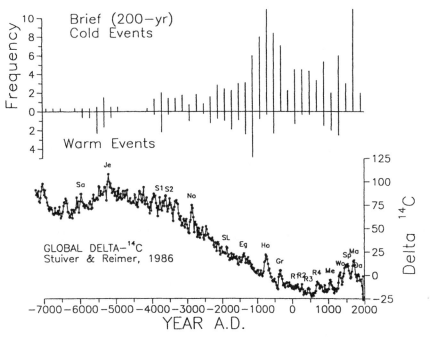

Fig. 2. Comparison of the frequency of brief (< 200 yr) climatic events with dendrochronologic record of ^{14}C anomalies. Climate data are dated (calibrated if ^{14}C-dated) cold events (e.g., glacial advance and lower tree line) and warm events (e.g., reduced sea ice and higher tree line); plotted against the mid-period of the 200 yr interval. The cold and warm events are compiled from various published sources (available from the author upon request), and include proxy indicators such as (cold) glacial advances, lake high-stands, and tree line retreats; versus (warm) glacier retreats, tree-ring maxima (boreal regions), and oxygen isotope maxima. Delta ^{14}C curve is after Stuiver and Reimer (1986). Labeled anomalies follow Eddy (1977), Landscheidt (1987), and Davis *et al.* (1992). Ma = Maunder, Sp = Sporer, Wo = Wolf, Me = Medieval, R1–4 = Roman, Gr = Greek, Ho = Homeric, Eg = Egyptian, Nd = Noachan deluge, S1–2 = Sumarian, Je = Jerico, Sa = Sahelan. Note coincidence of peak frequencies of cold events with large isotope anomalies (e.g., Homeric and Greek, Sporer and Maunder); and tendency for warm-event peaks to proceed cold-event peaks.

Stine, 1990). In general, the large anomalies – at 2200 yr intervals – correlate with increased frequency of cold indicators (e.g., glacial advance and lower tree line), and they are preceded by increased frequency of warm indicators (e.g., reduced sea ice and higher tree line).

The ages of the climatic events in Figure 2 primarily are determined by radio-carbon dates, which have age uncertainties greater than the length of the individual cosmogenic anomalies (Figure 2), so more precise correlation of climate and solar variability cannot generally be made with such ^{14}C-dated indicators. However, high-precision ^{14}C dating, accelerator ^{14}C dating, and curve matching permit more precise correlation under favorable circumstances (Van Geel and Mook, 1989).

In temperate regions, positive isotope anomalies are associated with cold, wet climate (Figure 2); and negative anomalies are linked with warm, dry climate, such as the MWP (Eddy, 1977; Hood and Jirkowic, 1990). However, in areas

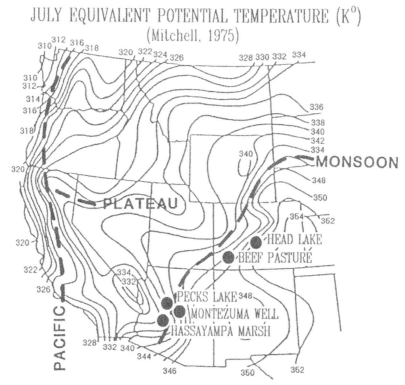

Fig. 3. Map of equivalent potential temperature of the western U.S.A. (after Mitchell, 1976) for July, with summer air mass boundaries, and location of pollen sites. Summer precipitation, originating in the Gulf of Mexico and in the western Pacific, is greatest (> 50% annual average) south of the monsoon airmass boundary, and reaches its northern limit at the plateau boundary.

with monsoonal climate, positive isotope anomalies may correlate with cool, *dry* events. For example, a dry period ca. 7000 yr B.P. in Sub-Saharan Africa (Gillespie *et al.*, 1983; Gasse *et al.*, 1990; Street-Perrott and Perrott, 1990) and China (Fang, 1991) correlates with a positive ^{14}C anomaly dated 7120–6840 B.C. (Stuiver and Braziunas, 1988). This regional difference may result from the effects of global cooling on the thermal contrast between tropical oceans and land, a primary driving mechanism for monsoonal circulation. If so, negative isotope anomalies (warming climate) should correlate with wet climate in areas with monsoonal climate.

1.4. *Climate of the American Southwest*

Today, the American Southwest receives up to half of its precipitation as summer monsoons, with the relative importance of winter precipitation increasing northward. The climatic patterns of western North America primarily are delimited by topographic features. Mitchell (1976) established air-mass boundaries based on the clustering of isolines of equivalent potential temperature (Figure 3). Winter

(January) boundaries are sharpest along the east slope of the Rocky Mountains, and along the Colorado River valley. Summer (July) boundaries are sharpest along the Pacific coast and the Colorado River valey. In both seasons a weak, 'plateau' boundary begins in the gap between the Cascade and Sierra Nevada mountains (41° N × 122° W), crosses the northern Great Basin along the Humboldt River valley and Great Salt Lake Basin, and ends near the Rawlins Gap of Wyoming (42° N × 107° W).

Winter precipitation, originating as frontal cyclonic storms, predominates north of the 'plateau' boundary and west of the Sierra Nevada. Summer precipitation, originating in the Gulf of Mexico and in the Western Pacific, is greatest south of the summer airmass boundary, and reaches its northern limit at the plateau boundary (Pyke, 1972). Recent climatological analyses by Tang and Reiter (1984) demonstrate that the highlands of western North America function in the same way as the Tibetan plateau to influence summer and winter atmospheric circulation.

Although the North American highlands are lower than the Tibetan plateau a strong anticyclonic shear line comparable to that over Sinkiang on the Tibetan Plateau forms in summer along the 'plateau' climatic boundary. The shear line is the northern boundary of the strong convection in summer (Tang and Reiter, 1984), and is the boundary between wind streamlines in winter (Mitchell, 1976, Figure 3). In summer, moist air is drawn northward in 2 low-level jets from the Gulf of California and Gulf of Mexico (Tang and Reiter, 1984).

1.5. *Human Activities and Rapid Climate Change*

Historic analogs for climatic change, such as the MWP and Little Ice Age, are valuable records of the effects of rapid climate change on human activities in Europe (Serre-Bachet, 1994) and China (Zhang and Mang, in press).

Rapid climatic change also may have influenced the behavior of prehistoric humans. Traditional indicators of rapid climatic change include the recurrence horizons (cf. *Grenzhorizont* or *rekurrens-ytor*) produced by drying (dark layers) followed by flooding (light layers) of peat bogs (Brooks, 1949; Lamb, 1977). These have been shown to vary in age from place to place (Mitchell, 1956). Human activities were influenced by some of these dry/wet events. They built corduroy bridges over humified peat when the bogs flooded, and these bridges later were buried by rapid growth of the bog. Godwin and Willis (1959) dated 4 of these corduroy 'trackways' in the Somerset region of England. The calibrated ages of these brief wet periods are coincident with cosmogenic isotope anomalies: 3700–3400 B.C., 700–500 B.C., 350–100 B.C., and A.D. 50–100 (Figure 4). Tree-ring methods have been used to date these trackways more precisely. Schmidt and Gruhle (1988) have dated the construction of trackways in southern Germany to 717–713 B.C., and Hillam *et al.* (1990) have dated the 'Sweet Track' in southwestern England to 3807 B.C.

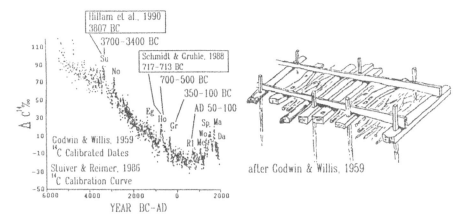

Fig. 4. A comparison of tree-ring and ^{14}C dated corduroy bridges from European bogs with global ^{14}C isotopic anomalies. References in boxes are tree-ring dates, others are calibrated ^{14}C dates. Corduroy 'trackway' right an archeological reconstruction showing longitudinal anchor planks that kept cross pieces from floating.

2. The Medieval Warm Period in the Southwest

2.1. *Southwestern Archeology and the Medieval Warm Period*

The Medieval Warm Period appears to have influenced human activities in the American Southwest. The prehistoric cultural groups of the Southwest: Anasazi, Mogollon, Hohokam, and Patayan, reached population and stylistic zeniths at the time of MWP (Cordell, 1984). All four groups lived within the Arizona monsoonal boundary (Figure 3). These were agricultural people who relied on dryland cultivation of corn, beans, squash, and cotton. The chronologies for the Patayan and Hohokam are less precise than that of the Mogollon and Anasazi, whose dwellings are dated by tree rings. Each group began slow population growth at about 500 A.D., reaching a climax ca. 1100–1200 A.D. and had begun major decline before A.D. 1400.

On the Colorado Plateau, U.S.A., the population density of the Anasazi people was greatest from ca. A.D. 1050–1300 (Figure 5), with the final population decline A.D. 1475 (Dean *et al.*, 1985). Major population shifts and abandonments at A.D. 1150 and A.D. 1275–1300 correlate with droughts (e.g., A.D. 1276–1299, the 'Great Drought' of A. E. Douglass [1935]). There is local variability in the archeological chronologies. At Hay Hollow on the Mogollon Rim of Arizona, a gradual population decline began ca. A.D. 1150 (Hevly, 1981); but in the Dolores Project Area (2000–2300 m; Petersen, 1988), the population peaked between A.D. 850 and 950 (Figure 5).

In recognition of the human value of climatic analogs, I will evaluate the association of isotope anomalies and climate for both the general Anasazi population

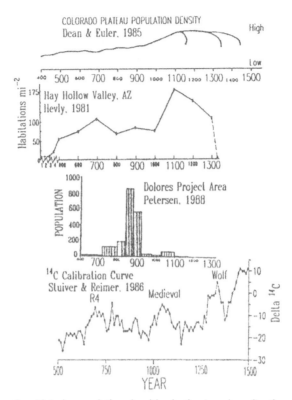

Fig. 5. Comparison of prehistoric population densities in the American Southwest with the global record of ^{14}C isotope anomalies.

maximum (A.D. 1050–1300; Dean *et al.*, 1985), and for the earlier population maximum of the Dolores Project Area (A.D. 850–950; Petersen, 1988). This 'Southwestern Warm Period' encompasses the European 'Medieval Warm Period', and spans two ^{14}C 'maxima' A.D. 700–1050 and A.D. 1050–1335 (Figure 5).

2.2. *Southwestern Paleoclimate during the Medieval Warm Period*

Two recent studies are excellent examples of the association of climate change and cosmogenic anomalies, and of climate during the Southwestern Warm Period. One is dendroclimatic, the other is palynological.

2.3. *Sierra Nevada Climate and Cosmogenic Isotope Anomalies*

Graumlich (1993) uses tree-ring data to reconstruct temperature and precipitation of the Southern Sierra Nevada to A.D. 800. The precipitation reconstructions from tree-ring data show no long-term trends, but the smoothed temperature estimates record sharp minima A.D. 1069, 1335, and 1524 that closely match the Medieval, Wolf, and Sporer isotopic anomalies (Figure 5). The highest temperature

(A.D. 1111) is during the Medieval (cosmogenic) 'Maximum' (A.D. 1050–1335). However, the correspondence of temperature minima and isotopic anomalies in Graumlich's record is poor for the Maunder and later anomalies. This mismatch might result either from changes in the global carbon cycle (Jirikowic *et al.*, 1993), or from the climate during the Maunder minimum being outside the tree-ring calibration envelope.

2.4. *Pollen Analysis and Rapid Climate Change*

Pollen analysis is a primary source of paleoenvironmental information for terrestrial environments. However, palynologists traditionally have neglected rapid climatic change, and instead have emphasized long-term climatic change. For example the post-glacial portion of the European Environmental Sequence consists of the 2000-yr moisture sequence (Sub-Boreal, Sub-Atlantic etc.) superimposed on the 10,000-yr thermal sequence (Iversen, 1973; Wendland and Bryson, 1974). Wright (1977) has emphasized long-term climate change for the upper midwest of North America; and Davis (1984) has postulated direct solar forcing of vegetation change in the northern Great Basin of Milankovitch periodicity (precession cycle, 19, 12 ky).

Despite the general neglect of rapid change, the palynologist's focus on long-term environmental change is primarily a matter of emphasis rather than a physical limitation on the technique of pollen analysis. Palynological study of vegetation change is limited by sedimentary processes; the temporal resolution is a function of the sediment accumulation rate, the sediment mixing rate, sediment deterioration, and sediment compaction (Hakanson and Jansson, 1983). For closely-spaced samples, deterioration and compaction are nearly uniform among samples, so the ratio of accumulation to mixing is of primary importance. In basins with very rapid sedimentation (accumulation >> mixing), or with very little mixing (annually-laminated sediments), pollen changes can be studied on an annual (MacDonald *et al.*, 1991) or seasonal (Tippett, 1964) scale.

2.5. *The Dolores River Project Area*

The most detailed palynological study of the climate during the Southwestern Warm Period is by Petersen (1988; 1994), who has compiled palynological and tree ring evidence for climatic change in the Dolores Project Area of the Colorado Plateau. Petersen uses pollen ratios of spruce/pine as an index of soil moisture, and the accumulation rate of pinyon pine an an index of summer precipitation. Both indicators are from the sediments from Beef Pasture, Colorado (Figure 3). Summer precipitation was greatest (> 800 pinyon/cm^2/yr) from A.D. 780–A.D. 1130, with a brief decline at A.D. 980. Soil moisture was greatest (spruce/pine > 0.5) before A.D. 800, from A.D. 1025–1110, and after A.D. 1325.

Ring width indices from Alamagre Mountain, Colorado, provide an index of the length of the growing season. Indices are greatest (> 110) at A.D. 630, 830, and 970;

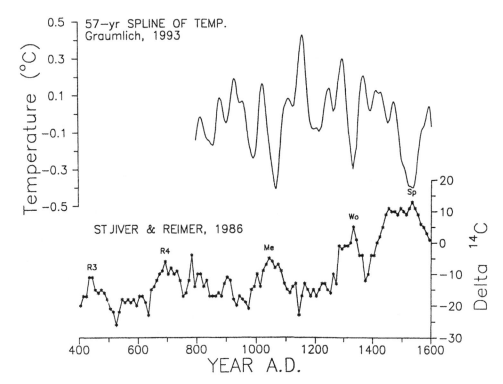

Fig. 6. Comparison of climate of the Sierra Nevada, California, U.S.A. (Graumlich, 1993) with the global record of ^{14}C isotope anomalies.

and from A.D. 1110–1190 (cf. maximum temperature in Sierra Nevada A.D. 1111, Figure 6). Peak population density at the Dolores Project Area coincides with high summer moisture and low soil moisture peak at A.D. 1025–1100 coincides with the Medieval isotope anomaly, but summer precipitation maximum shows little relationship to the ^{14}C curve. Maximum pinyon pollen/cm^2/yr spans the Roman IV maximum and Medieval minimum (Figure 7).

In summary, climate reconstructions for the Southwest show some correspondence with the cosmogenic isotope record (Figures 5, 6) during the Southwestern Warm Period (A.D. 700–1350), and indicate maximum warmth ca. A.D. 1100–1200. In southwestern Colorado summer precipitation was greatest before from A.D. 700–1150 (Figure 7).

3. Palynology of Selected Southwestern Sites

To further evaluate climatic change during the Southwestern Warm Period, I will compare Petersen's (1988) reconstructions (Figure 7) with palynological studies from 4 other sites south of the summer monsoon boundary (Figure 3). The sites were not originally selected for the study of rapid climate change, so their sampling

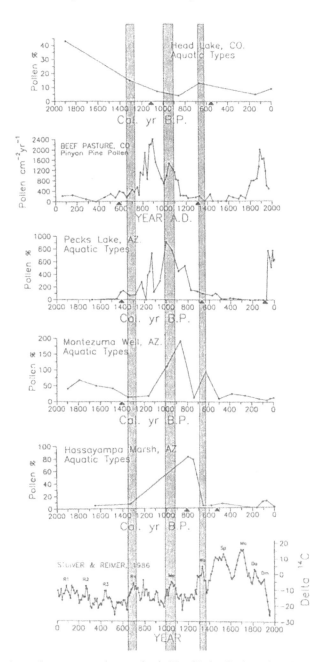

Fig. 7. Comparison of reconstructed water depth (Head Lake, Pecks Lake, Hassayampa Marsh) and water quality (Montezuma Well) with summer precipitation (Beef Pasture) and the global record of ^{14}C isotope anomalies. Water depth and quality based on percentages of aquatic pollen and spores. Following convention, the percentages are calculated with a divisor of the sum of upland pollen, so percentages may exceed 100%. Sites are arranged by latitude from north (Head Lake) to south (Hassayampa Marsh). Triangles mark age of radiocarbon dates for each site. Shaded bars are times of cosmogenic isotope anomaly.

TABLE I: Comparison of dating, sedimentation rates, and sampling frequency for Southwestern pollen-analytic sites in studying the Medieval Warm Period.

Site	Bracketing [14]C Dates yr B.P.	Sedimentation Rate cm/yr	Sample cm	Frequency yr
Head Lake, CO 37°33′ 105°30′	670 ± 60 – 1430 ± 60	0.013	3	280
Beef Pasture, CO 37°25′ 108°09′	780 ± 90 – 1540 ± 80	0.099	2.5	28
Pecks Lake, AZ 37°47′ 112°02′	650 ± 180 – 1485 ± 60	0.176	10	64
Montezuma Well, AZ 34°39′ 112°45′	surface – 1526 ± 50	0.167	20	76
Hassayampa, AZ 33°55′ 112°42′	485 ± 60 – 955 ± 60	0.179	28	157

frequency is relatively low (Table I). However, each site has ^{14}C dates during the Southwestern Warm Period. To compare these records with the isotopic chronologies, the ^{14}C dates have been calibrated (Stuiver and Reimer, 1986). At each site, algae and the pollen of aquatic plants are used as paleoecological indicators of water depth and water chemistry.

Head Lake (Jodry *et al.*, 1989) is the northernmost site herein compared (Figure 3). It is located east of Beef Pasture, in the bottom of the Alamosa Valley, Colorado (2300 m), in the rain shadow of surrounding mountains. The sedimentation rate is slow (0.013 cm yr^{-1}, Table I), but 9 bulk-sediment ^{14}C dates in the Holocene provide reliable chronologic control. The water-depth index for Head L. (Figure 7) is the sum of standing water indicators (*Pediastrum* [square root] + *Botryococcus* + Cyperaceae) minus wet ground indicators (*Salix*-fern spores).

Pecks Lake (1016 m) is in the Verde River Valley of central Arizona (Davis and Turner, 1987). The lake's sedimentation is more rapid than Beef Pasture, but its sampling frequency is less (Table I). The dates shown in Figure 7 are based on plant macrofossils, because the bulk-sediment ^{14}C dates for Pecks Lake are unreliable (Davis and Turner, 1987; Davis, unpublished). The water-depth index (Figure 7) for Pecks Lake is based on the results of historic flooding of the basin; it is riparian (*Fraxinus* × 2 + *Populus*) plus shallow water (Cyperaceae/2 + Potamogeton + Typha-Sparganium/2 + Typha *latifolia*/2) plus submerged acquatic (*Myriophyllum* × 2).

Montezuma Well (1125 m) also is in the Verde River Valley (Figure 3). The site occupies a sink hole in Verde Formation limestone, and is fed by basal springs. Due to well-documented dating problems (Hevly, 1974; Davis and Shafer, 1992), all of its dates are AMS dates for terrestrial macrofossils. The water-depth index (Figure 7) for Montezuma Well includes riparian (*Fraxinus*) and low-salinity, clearwater taxa (*Pediastrum + Spirogyra*/2).

Hassayampa Marsh (Davis, 1990) is the lowest (606 m) and southernmost of the sites. Its sedimentation rate is very rapid, and it is well dated, but relatively few pollen samples have been processed so far (Table I). The water-depth index (Figure 7) for Hassayampa Marsh includes the two most abundant aquatic taxa (Cyperaceae + *Spirogyra*).

3.1. *Results*

The 3 southern sites indicate greatest water-depth or freshness coincident with Beef Pasture's indications of increased summer precipitation (Figure 7). The peaks are abrupt, large and prolonged from A.D. 700–1350. No other events in the last 2000 years rival the water-depth values for this period, except the artificial flooding (> 3 m) of Pecks Lake ca. A.D. 1910 (Davis and Turner, 1987). In contrast, Head Lake records low water depth during this time. From Beef Pasture southward, the date of peak values appears to be progressively younger.

3.2. *Discussion*

The reconstructions of water depth (Pecks Lake, Hassayampa Marsh) and water quality (Montezuma Well) support Petersen's (1988, 1994) findings of greater (summer) moisture during the Southwestern Warm Period. This high-lake interval also is documented at the Salton Sea where the 'third lacustrine unit' is dated 1187–1312 cal. A.D. (Waters, 1989). The lake-level data clearly indicate increased moisture coincident with the florescence of the Anasazi, Mogollon, Hohokam, and Patayan cultures.

However, the relationship of climate to solar activity is uncertain, due to inadequate ^{14}C dating. Although the moisture peak is associated with cosmogenic maxima as expected for a monsoonal region, the effect of the Medieval minimum is unclear. The Beef Pasture highest pinyon pollen influx is during the Roman IV-Medieval maximum, and the Hassayampa Marsh aquatic pollen peak is during the Medieval-Wolf maximum (485 ± 60 yr B.P., 1144–1427 cal yr B.P.). Pecks Lake and Beef Pasture record two peaks separated by a trough. The lesser peak at Beef Pasture and the major peak at Pecks Lake coincide with the Medieval minimum (Figure 7). Although this could indicate an inconsistent (or complex) relationship between precipitation and the ^{14}C minima, the age of the peaks is inferred by interpolation between dated horizons (Figure 7). Changes of sedimentation rates by 48% (Beef Pasture) or 25% (Pecks Lake) could align the troughs with the Medieval minimum and peaks with the 2 maxima.

These chronologic problems can be overcome with further ^{14}C dating. Montezuma Well needs several ^{14}C dates younger than 1526 ± 60 yr B.P. At least two more dates between 1400 and 700 yr B.P. are needed at Pecks Lake and Beef Pasture, and a date >955 ± 60 yr B.P. and many more pollen samples are needed for Hassayampa Marsh.

4. Conclusions

The pollen analysis of three sites (Pecks Lake, Hassayampa Marsh, and Montezuma Well) within the Arizona Monsoon Boundary indicate increased lake level from A.D. 700–1350, supporting Petersen's (1988, 1994) indication of summer precipitation greater than today. Inadequate dating makes detailed comparison with the ^{14}C isotope record inconclusive, but the records all indicate maximum summer moisture coincident with maximum population density of Southwestern prehistoric people.

This association of a *negative* cosmogenic isotope anomaly with *increased* moisture, together with the association of positive anomalies with dry events in other monsoonally-influenced regions (Gillespie *et al.*, 1983; Gasse *et al.*, 1990; Street-Perrott and Perrott, 1990; Fang, 1991), suggests a positive influence of global temperature on monsoonal precipitation. This effect is opposite to that for regions dominated by winter cyclonic storms (Eddy, 1977; Schmidt and Gruhle, 1988; Davis, 1992).

The aridity during the MWP apparent at Head Lake (Figure 7), the northernmost site in this study, is consistent with climatic records elsewhere in western U.S., north of the Monsoon Boundary (e.g. the eastern Sierra Nevada (Stine, 1990). Together with the apparent southward increase in the age of maximum lake level, these data might indicate the influence of rapid warming and cooling on atmospheric circulation. Cosmogenic 'maxima' may be associated with enhanced monsoonal circulation, and isotope 'minima' with increased frequency of frontal cyclonic storms.

Acknowledgements

Financial support for the study of Head Lake was provided by the Smithsonian Institution through Dennis Stanford, and by the National Science Foundation (ATM 8619467). Analysis of Montezuma Well was supported by University of Arizona Grant 424002, and NSF grants ATM-8619467 and SES-9009974. Analysis of Hassayampa Marsh was supported by the Arizona Nature Conservancy. Lisa Graumlich and Kenneth Petersen kindly provided their data for inclusion in this study. The radiocarbon anomaly data are from file ATM20.14C, part of the CALIB program distributed by the University of Washington radiocarbon laboratory.

References

Barnola, J. M., Raynaud, D., Korotkevich, T. S., and Lorus C.: 1987, 'Vostock Ice Core Provides 160,000-Year Record of Atmospheric CO₂', *Nature* **329**, 408–413.
Beer, J., Siegenthaler, U., Bonani, G., Finkel, R. C., Oeschger, H., Suter, M., and Wolfli, W.: 1988, 'Information on Past Solar Activity and Geomagnetism from ^{10}Be in the Camp Century Ice Core', *Nature* **331**, 675–679.
Brooks, C. E. P.: 1949, *Climate through the Ages*, H. and J. Pillans and Wilson, Printers, Edinburgh, p. 439.

Bryson, R. A.: 1988, 'Late Quaternary Volcanic Modulation of Milankovitch Climate Forcing', *Theor. Appl. Climatol.* **39**: 115–125.

Cordell, L. S.: 1984, *Prehistory of the Southwest*, Academic Press, New York, p. 409.

Damon, P. E.: 1968, 'Radiocarbon and Climate (A Comment on a Paper by H. Suess)', *Meteorol. Monogr.* **8**, 151–154.

Damon, P. E. and Sonett, C. P.: 1991, 'Solar and Terrestrial Components of the Atmospheric [14]C Variation Spectrum', in Sonnett, C. P., Giampapa, M. S. (eds.), *The Sun in Time*, University of Arizona Press, pp. 360–388.

Davis, O. K.: 1984, 'Multiple Thermal Maxima during the Holocene', *Science* **255**, 617–619.

Davis, O. K.: 1990, 'Pollen Analysis of Hassayampa Preserve, Maricopa Co, Arizona', The Nature Conservancy, Hassayampa River Preserve, Box 1162, Wickenburg, Arizona.

Davis, O. K.: 1992, 'Rapid Climatic Change in Coastal Southern California Inferred from Pollen Analysis of San Joaquin Marsh', *Quat. Res.* **37**, 89–100.

Davis, O. K. and Turner, R. M.: 1987, 'Palynological Evidence for the Historic Expansion of Juniper and Desert Shrubs Resulting from Human Disturbance in Arizona, U.S.A.', *Rev. Palaeobot. Palynol.* **49**, 177–193.

Davis, O. K. and Shafer, D. S.: 1992, 'An Early-Holocene Maximum for the Arizona Monsoon Recorded at Montezuma Well, Central Arizona', *Palaeogeogr. Palaeoclimatol. Palaeocol.* **92**, 107–119.

Davis, O. K., Jirikowic, J. L., and Kalin, R. M.: 1992, 'The Radiocarbon Record of Solar Variability and Holocene Climatic Change in Coastal Southern California', Proceedings 8th PACLIM Workshop, *Calif. Dept. Water Res. Interagency Ecol. Stud. Progr. Techn. Rep.* **31**, 19–33.

Dean, J. S., Euler, R. C., Gumerman, G. J., Plog, F., Hevly, R. H., and Karlstrom, N. V.: 1985, 'Human Behavior Demography and Paleoenvironment on the Colorado Plateau', *Amer. Antiq.* **50**, 537–554.

Douglass, A. E.: 1935, 'Dating Pueblo Bonito and Other Ruins of the Southwest', *Nat. Geogr. Soc. Pueblo Bonito Ser.* **1**, 1–74.

Eddy, J. A.: 1976, 'The Maunder Minimum', *Science* **192**, 1189–1202.

Eddy, J. A.: 1977, 'Climate and the Changing Sun', *Climate Change* **1**, 173–190.

Fang, J.: 1991, 'Lake Evolution during the Past 30,000 Years in China, and Its Implications for Environmental Change', *Quat. Res.* **36**, 37–60.

Gasse, F., Tehet, R., Durand, A., Gibert, E., and Fonts, J–C.: 1990, 'The Arid-Humid Transition in the Sahara and the Sahel during the Last Deglaciation', *Nature* **346**, 141–146.

Van Geel, B. and Mook, W. G.: 1989, 'High-Resolution ^{14}C dating of Organic Deposits Using Natural atmospheric [14]C Variations', *Radiocarbon* **31**, 151–155.

Gillespie, R., Street-Perrot, F. A., and Switsur, R.: 1983, 'Post-Glacial Arid Episodes in Ethiopia Have Implications for Climate Prediction', *Nature* **306**, 680–683.

Godwin, H. and Willis, E. H.: 1959, 'Radiocarbon Dating of Prehistoric Wooden Trackways', *Nature* **184**, 490–491.

Graumlich, L. J.: 1993, 'A 1000-Year Record of Temperature and Precipitation in the Sierra Nevada', *Quat. Res.* **39**, 249–255.

Grove, J. M.: 1988, *The Little Ice Age*, Methuen, London, p. 498.

Hakanson, L. and Jansson, M.: 1983, *Principles of Lake Sedimentation*, Springer-Verlag, New York, p. 316.

Hevly, R. H.: 1974, 'Recent Paleoenvironments and Geological History At Montezuma Well', *J. Ariz. Acad. Sci.* **9**, 66–75.

Hevly, R. H.: 1981, 'Pollen Production, Transport and Preservation: Potentials and Limitation in Archaeological Palynology', *J. Ethnobiol.* **1**, 39–55.

Hilliam, J., Groves, C. M., Brown, D. M., Baillie, M. G. L., Coles, J. M., and Coles, B. J.: 1990, 'Dendrochronology of the English Neolithic', *Antiquity* **64**, 210–220.

Hood, L. L. and Jirikowic, J. L.: 1990, 'Recurring Variations of Probable Solar Origin', *Geophys. Res. Lett.* **17**, 85–88.

Iversen, J.: 1973, 'The Development of Denmark's Nature since the Last Glacial', *Danm. Geol. Undersøg.* **5** (7C).

Jirikowic, J. L., Kalin, R. M., and Davis, O. K.: 1993, 'Tree-Ring ^{14}C as an Indicator of Climate

Change', Proceedings, Chapman Conference, Jackson Hole, Wyoming, AGU, *Geophys. Monogr.* **78**, 353–366.

Jodry, M. A., Shafer, D. S., Stanford, D. J., and Davis, O. K.: 1989, 'Late Quaternary Environments and Human Adaptation in the San Luis Valley, South-Central Colorado', Colorado Ground-Water Association Guidebook, 8th Annual Field Trip, pp. 189–208.

Kromer, B., Rhein, M., Bruns, M., Schoch-Fischer, H., Munnich, K. O., Stuiver, M., and Becker, B.: 1986, 'Radiocarbon Calibration Data for the 6th to 8th Millennia B.C.', *Radiocarbon* **28** (2B), 954–960.

Lamb, H. H.: 1977, *Climate History and the Future*, Princeton University Press, p. 835.

Landscheidt, T.: 1987, 'Long-Range Forecasts of Solar Cycles and Climate Change', in Rampino, M. R., Sanders, J. E., Newman, W. S., and Konigsson, L. K. (eds.), *Climate History, Periodicity, and Predictability*, Van Nostrand Reinhold, New York, pp. 421–445.

Lean, J.: 1991, 'An Estimate of the Sun's Photon Output during the Maunder Minimum', Abstract SH41A-3 Spring AGU Meeting, p. 224.

Linick, T. W., Suess, H. E., and Becker, B.: 1985, 'La Jolla Measurements of Radiocarbon in South German Oak Tree-Ring Chronologies', *Radiocarbon* **27**, 20–32.

Linick, T. W., Long, A., Damon, P. E., and Ferguson, C. W.: 1986, 'High-Precision Radiocarbon Dating of Bristlecone Pine from 6554 to 5350 B.C.', *Radiocarbon* **28** (2B), 943–953.

MacDonald, G. M., Larsen, C. P. S., Sceicz, J. M., and Moser, K. A.: 1991, 'The Reconstruction of Boreal Forest Fire History from Lake Sediments: A Comparison of Charcoal, Pollen, Sedimentological, and Geochemical Indices', *Quat. Sci. Rev.* **10**, 53–72.

Mitchell, G. F.: 1956, 'Post-Boreal Pollen Diagrams from Irish Raised-Bogs', *Proc. Royal Irish Acad.* **57**, 185–251.

Mitchell, V. L.: 1976, 'The Regionalization of Climate in the Western United States', *J. Appl. Meteorol.* **15**, 920–927.

Pearson, G. W. and Stuiver, M.: 1986, 'High-Precision Calibration of the Radiocarbon Time Scale, 500–2500 B.C.', *Radiocarbon* **28** (2B), 839–862.

Pearson, G. W., Pilcher, J. R., Baillie, M. G. L., Corbett, D. M., and Qua, F.: 1986, 'High-Precision ^{14}C Measurement of Irish Oaks to Show the Natural ^{14}C Variations from A.D. 1840–5210 B.C.', *Radiocarbon* **28** (2B), 911–934.

Petersen, K. L.: 1988, 'Climate and the Dolores River Anasazi', *Univ. Utah Anthropol. Pap.* **113**, 152 pp.

Petersen, K. L.: 1994 'A Warm and Wet Little Climatic Optimum and a Cold and Dry Little Ice Age in the Southern Rocky Mountains, U.S.A.', in Hughes, M. K. and Diaz, H. F. (eds.), *Clim. Change* **26**, 243–269, (this issue).

Post, W. M., Peng, T.-H., Emanuel, W. R., King, A. W., Dale, V. H., and DeAngelis, D. L.: 1990, 'The Global Carbon Cycle', *Amer. Sci.* **78**, 310–326.

Pyke, C. B.: 1972, 'Some Meteorological Aspects of the Seasonal Distribution of Precipitation in the Western United States and Baja California', *Univ. Calif. Water Resour. Cent. Contrib.* **139**, 205 pp.

Raisbeck, G. M., Yiou, F., Jouzel, J., and Petit, J. R.: 1990, '^{10}Be and δ^2H in Polar Ice Cores as a Probe of the Solar Variability Influence on Climate', *Philos. Trans. R. Soc. London* **A330**, 463–470.

Schiffer, M. B.: 1982, 'Hohokam Chronology: An Essay on History and Method', in McGuire, R. H. and Schiffer, M. B. (eds.), *Hohokam and Payatan: Prehistory of Southwestern Arizona*, Academic Press, New York, pp. 299–344.

Schmidt, B. and Gruhle, W.: 1988, 'Klima, Radiokohlenstoffgehalt und Dendrochronologie', *Naturwissensch. Rundsch.* **41**, 177–182.

Serre-Bachet, F.: 1994, 'Middle Ages Temperature Reconstructions in Europe: A Focus on Northeastern Italy', in Hughes, M. K. and Diaz, H. F. (eds.), *Clim. Change* **26**, 213–224, (this issue).

Smith, J. B. and Tirpak, D. A.: 1989, *The Potential Effects of Global Climate Change on the United States*, U.S. Environm. Protect. Agency, Washington, D.C. EPA–230–05–89.

Stine, S.: 1990, 'Late Holocene Fluctuations of Mono Lake, Eastern California', *Palaeogeogr. Palaeoclimatol. Palaeoecol.* **78**, 333–381.

Street-Perrott, F. A. and Perrott, R. A.: 1990, 'Abrupt Climate Fluctuations in the Tropics: The Influence of Atlantic Ocean Circulation', *Nature* **343**, 607–612.

Stuiver, M.: 1965, 'Carbon-14 Content of 18th and 19th Century Wood: Variations Correlated with Sunspot Activity', *Science* **149**, 533–537.

Stuiver, M. and Pearson, G. W.: 1986, 'High-Precision Calibration of the Radiocarbon Time Scale, A.D. 1950–500 B.C.', *Radiocarbon* **28** (2B), 805–838.

Stuiver, M. and Reimer, P. J.: 1986, 'A Computer Program for Radiocarbon Age Calibration', *Radiocarbon* **28**, 1022–1030.

Stuiver, M., Braziunas, T. F., Becker, B., and Kromer, B.: 1991, 'Climatic, Solar, Oceanic, and Geomagnetic Influences on Late-Glacial and Holocene Atmospheric $^{14}C/^{12}C$ Change', *Quat. Res.* **35**, 1–24.

Stuiver, M., Kromer, B., Becker, B., and Ferguson, C. W.: 1986, 'Radiocarbon Age Calibration back to 13,300 years B.P. and the ^{14}C Age Matching of the German Oak and U.S. Bristlecone Pine Chronologies', *Radiocarbon* **28** (2B), 969–979.

Suess, H. E.: 1965, 'Secular Variations of the Cosmic-Ray-Produced Carbon 14 in the Atmosphere and Their Interpretations', *J. Geophys. Res.* **70**, 5937–5952.

Tang, M. and Reiter, E. R.: 1984, 'Plateau Monsoons of the Northern Hemisphere: A Comparison between North America and Tibet', *Month. Wea. Rev.* **112**, 617–637.

Tippett, R.: 1964, 'An Investigation into the Nature of the Layering of Deep-Water Sediments in Two Eastern Ontario Lakes', *Can. J. Bot.* **42**, 1693–1704.

de Vries, H. L.: 1958, 'Variation in Concentration of Radiocarbon with Time and Location on Earth', *Koninkl. Nederl. Akad. Wetensch., Proc.* **B61**, 94–102.

Waters, M. R.: 1989, 'Late Quaternary Lacustrine History and Paleoclimatic Significance of Pluvial Lake Cochise, Southeastern Arizona', *Quat. Res.* **32**, 1–12.

Wendland, W. M. and Bryson, R. A.: 1974, 'Dating Climatic Episodes of the Holocene', *Quat. Res.* **4**, 9–24.

Wright, H. E.: 1977, 'Quaternary Vegetation History – Some Comparisons between Europe and America', *Ann. Rev. Earth Plant Sci.* **5**, 123–158.

Zhang, De'er: 'Evidence for the Existence of the Medieval Warm Period in China', in Hughes, M. K. and Diaz, H. F. (eds.), *Clim. Change* **26**, 289–297 (this issue).

(Received 22 September, 1992; in revised form 1 November, 1993)

EVIDENCE FOR THE EXISTENCE OF THE MEDIEVAL WARM PERIOD IN CHINA

ZHANG DE'ER

Chinese Academy of Meteorological Sciences, Baishiqiaolu No. 46 Beijing 100081, China

Abstract. The collected documentary records of the cultivation of citrus trees and *Boehmeria nivea* (a perennial herb) have been used to produce distribution maps of these plants for the eighth, twelfth and thirteenth centuries A.D. The northern boundary of citrus and *Boehmeria nivea* cultivation in the thirteenth century lay to the north of the modern distribution. During the last 1000 years, the thirteenth-century boundary was the northernmost. This indicates that this was the warmest time in that period. On the basis of knowledge of the climatic conditions required for planting these species, it can be estimated that the annual mean temperature in south Henan Province in the thirteenth century was 0.9–1.0 °C higher than at present. A new set of data for the latest snowfall date in Hangzhou from A.D. 1131 to 1264 indicates that this cannot be considered a cold period, as previously believed.

1. Introduction

Since the thermal optimum of the Holocene, the general trend of temperature in China has been a decline with secondary fluctuations of cooling and warming. The cooling stage, namely the Little Ice Age, has been discussed in more detail elsewhere (Grove, 1988; Zhang, 1991, 1992), as have the warming stages in the Han Dynasty (second century B.C.) and the Tang Dynasty (seventh to ninth centuries) (Chu, 1973). As for the Medieval Warm Period, nominally assigned to A.D. 900 to 1300, its existence in China has also been established by reference to contemporary documents.

There exist numerous records of the cultivation of citrus trees and the harvest of *Boehmeria nivea* (a perennial herb) in old Chinese documents. These records reflect the changes in where they were planted at different times. Both citrus and *Boehmeria nivea* are subtropical plants, whose growth and yield depend so closely on thermal conditions that we can adopt their planting boundary as a good indicator of temperature variations. We may also consider the period in which they were grown furthest north as the warmest period. Knowledge of modern climate limitations of their successful cultivation and harvest can be used to infer historical climatic conditions as their northern boundary. Chu (1931, 1973) considered the period around A.D. 1200 to be cold. This conclusion was in part based on a set of data on snowfall dates in Hangzhou in the South Song Dynasty. Recently the author's work based on a recalculation of the calendar conversion from the original lunar calendar to a solar calendar for ancient original records, revealed that almost all of the recalculated dates are earlier than thought hitherto. Consequently, the previous conclusion should be revised. This issue will be discussed further below.

Climatic Change **26**: 289–297, 1994.
© 1994 *Kluwer Academic Publishers.*

2. Information Obtained from Ancient Records of Planting *Boehmeria nivea*

2.1. *The Historical Planting Distribution*

Boehmeria nivea is a tropical-subtropical perennial herb and has been used as a raw material for clothing since ancient times. It is very sensitive to temperature variation and would yield two to five crops a year depending on the thermal conditions. Chinese historical documents contain detailed descriptions of the planting and harvesting of this herb. In this paper, only descriptions extracted from the official documents of the Tang Dynasty (A.D. 618–907), the North Song Dynasty (A.D. 960–1127) and the Mongol-Yuan Dynasty (A.D. 1206–1368) are used to map planting of *Boehmeria nivea* (Figure 1).

For the eighth century, a total of 130 planting places, now located in 101 counties (Tan, 1982) were found in the 'New Tang Dynasty Book' (Ou Y.X., 1061). In the 'Song Dynasty History', 133 planting places, now located in 101 counties, are noted (Tan, 1982). These are shown for the years A.D. 1102–1105. For the Yuan Dynasty, the relevant records are obtained exclusively from an official agriculture direction book 'Nong Sang Ji Yao', compiled by 'Da Si Nong Si' (ministry of agriculture) on orders from the emperor to direct agricultural production for the whole country in A.D. 1264 and issued by government in A.D. 1273. In this book there are descriptions such as:

> "the *Boehmeria nivea* planted in Southern China previously can be planted in Henan province now (A.D. 1264)"

> "at present (A.D. 1264), around Chen county and Cai country... there can be three crops a year... the yield is 30 *Jin* (a weight unit equal to 0.5 kg) per *Mu* (a unit of area, equal to 0.0667 ha) and the price is 300 *Wen* (A monetary unit) per Jin" (Wang Pan, 1273)

This means that Chen county (i.e. Huiyang today) and Cai county (i.e. Runan today) were the northernmost boundary of *Boehmeria nivea*, and three crops a year were obtained there in the mid-thirteenth century. It is noteworthy that the *Boehmeria nivea* had become a new local production in these places in the mid-thirteenth century. This fact reflects that the warm, stable climate favored this production. Comparing these planting distributions for different dynasties, it can be seen that, during the Yuan Dynasty, the northern boundary lay to the north of that which existed during the Tang and the Song dynasties, and much further north than today (Figure 1).

2.2. *Inference of Temperature at the Northern Boundary of Boehmeria nivea in the Thirteenth Century*

The growth and harvest of *Boehmeria nivea* require certain temperature conditions. In general, this plant requires an accumulated temperature of 1400–1700 °C greater

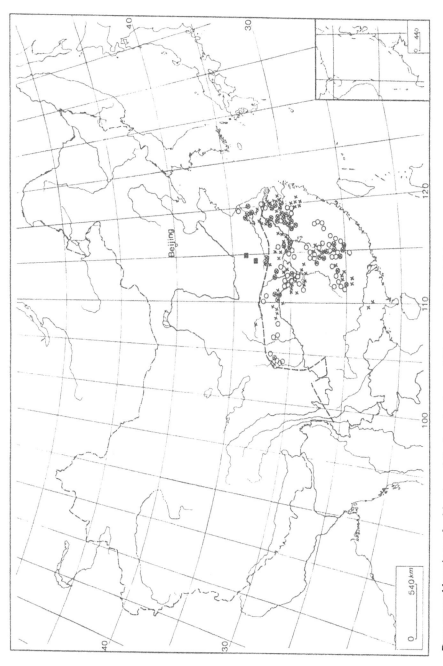

Fig. 1.   Documented locations of cultivation of *Boehmeria nivea* at different times: the eighth century (open circle), A.D. 1102–1105 (cross) and A.D. 1264 (closed square). The northern boundary of the favorable zone with 3 crops a year at present is shown as the broken line.

than 10 °C. The temperature range favorable to growth is 9–35 °C (Han, 1991). It
has two crops a year in the temperate zone and five crops a year in regions with
warmer, wetter conditions. The climatic regionalization of *Boehmeria nivea* is divided into four zones: the most, favorable zone, subfavorable zone and unfavorable
planting zone (Cheng, 1991). *Boehmeria nivea* can yield three crops a year only
in the favorable zone, where the annual temperature is about 15.5–16.5 °C, mean
monthly temperature in January is 1.5–3 °C and the soil minimum temperature at
5 cm depth in January is 3.2–4.5 °C (from agroclimatic data stored in the Chinese
Academy of Meteorological Sciences). In the thirteenth century, the northernmost
places with three crops a year were located in Huiyang and Runan. Based on meteorological records (1961–1980), the annual mean temperature at Huiyang and
Runan are 14.5 °C and 15.1 °C, mean January temperatures are 0.9 °C and 1.2 °C
and extreme minimum temperatures are −12.3 °C and −10.0 °C respectively. It is
warmer at Runan than Huiyang. Thus, we can infer the temperatures at Huiyang
in the thirteenth century by the lower limit of temperatures (15.5 °C, 1.5 °C) of
the favorable planting zone of *Boehmeria nivea*, and the inferred temperatures at
Runan should be greater than the lower limit value. Table I shows the inferred
temperature values at Huiyang and Runan, the northernmost places with three
crops a year in the thirteenth century. This indicates that annual temperatures in
the thirteenth century are nearly 1 °C higher than at present, and that the monthly
mean temperature in January was about 0.6 °C higher.

TABLE I: Comparison between the temperature value (°C) inferred in the thirteenth century (A)
and that measured at present (B)

Place	Annual Temp.			January temp.			Extr. Min. Temp.		
	A	B	A–B	A	B	A–B	A	B	A–B
Hui Yang	15.5	14.5	1.0	1.5	0.9	0.6	–	−12.3	–
Ru Nan	>15.5	15.1	–	>1.5	1.2	–	–	−10.0	–
Den Xian	>16.0	15.3	–	>2.0	1.7	–	>7.0	−9.8	–
Tang He	16.0	15.1	0.9	2.0	1.4	0.6	−7.0	−10.5	3.5

3. Information Obtained from Ancient Records of Planting Citrus

3.1. *The Historical Planting Distribution of Citrus*

Citrus trees prefer a warm and moist climate, and have a long history of cultivation
in China. From the 'New Tang Dynasty Book' (Ou Y. X., 1061), we have collected
the records of 145 locations where cultivated citrus trees grew. These locations
today span 91 counties (Figure 2) and show the actual distribution of citrus in the
eighth century. As for the Yuan Dynasty, a significant record was obtained from
the official agriculture direction book 'Non Sang Ji Yao', viz.: –

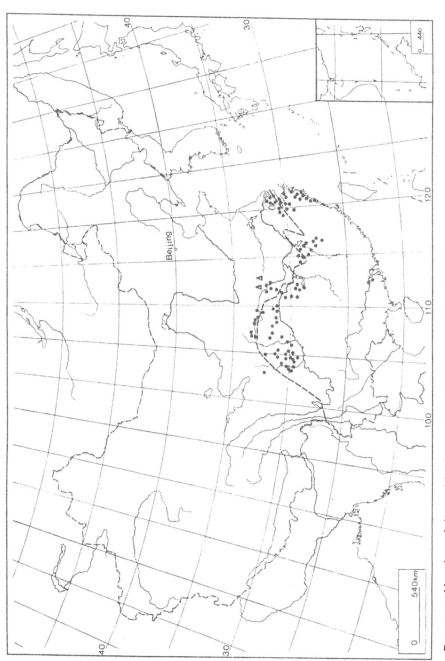

Fig. 2.   Documented locations of citrus cultivation around the eighth century (closed circle) and in A.D. 1264 (triangle), and the northern boundary of the subfavorable planting region at present (broken line).

"Orange is a new crop found in Xichuan, Tang county and Deng county. Citrus has been cultivated successfully in these places recently (A.D. 1264)" (Wang Pan, 1273)

This record reflects actual cultivation in the period around A.D. 1264. These places (i.e. Tanghe county and Deng county today) can be considered as the northernmost county of citrus cultivation at that time (Figure 2). It can be shown that the northern boundary of citrus cultivation in the mid-thirteenth century was north of that in the Tang Dynasty and even further north than the present limit (He Kang, 1985).

3.2. Inference of Temperature on the Northern Boundary of Citrus Trees in the Thirteenth Century

The cultivation of citrus trees is sensitive to thermal conditions, especially to the minimum. In general, it is considered necessary for successful citrus cultivation that the minimum temperature be above −5 °C. The study of climatic regionalization of citrus (Chinese Academy of Agriculture Science, 1980) based on present climatic data results in the designation of five cultivating zones with defined temperature indicators. The subfavorable cultivation zone is defined by annual mean temperatures over 16 °C, monthly mean temperatures 2–5 °C and mean extreme minimum temperatures greater than −7 °C. It is noteworthy that the recent northern boundary of the subfavorable zone (indicated in Figure 2 by the broken line) almost coincides with the present main citrus-producing area. According to such climatic indicators, we can consider temperature conditions on the northern boundary of the subfavorable zone as being: annual mean temperature is 16 °C, monthly mean temperature in January is 2 °C and mean extreme minimum temperature is −7°C. Therefore, supposing the temperature conditions of the northernmost citrus cultivation in the thirteenth century to be analogous to those of the northern boundary of the subfavorable zone today, the temperature of Tanghe county and Deng county in the mid-thirteenth can be inferred as shown in Table I. The indications are that the annual mean temperature in the mid-thirteenth century was 0.9 °C higher, the monthly mean January temperature was 0.6 °C higher, and the mean extreme minimum temperature was probably 3.5 °C higher than at present.

4. New Data on the Spring Snowfall Date in Hangzhou During the South Song Time (A.D. 1131–1264)

Records of spring snowfall dating from A.D. 1131 to 1264 in Hangzhou, the capital of the South Song Dynasty, have hitherto been used to define this as a period of cold conditions (Chu, 1973). Recently, we have obtained more records from other historical documents (shown in Table II and marked by *, **). We checked all of these original records and re-examined their calendar conversion from the lunar calendar to the solar calendar using new conversion books (Chen, 1962; Fang, 1987). We found that almost all the revised dates of spring snowfall are earlier

TABLE II: The latest date of spring snowfall in Hangzhou (A.D. 1131–1264)

Year	Date (revised)	Date (previous)	Difference (day)
1131	Apr. 4	Apr. 11	+7
1135	Mar. 16	Mar. 24	+7
1136	Mar. 8	Mar. 15	+7
1143	Apr. 12	Apr. 19	+7
1147	Mar. 5	Mar. 12	+7
1148	Mar. 6	Mar. 13	+7
1158	Apr. 6	Apr. 13	+7
1159	Mar. 4	Mar. 11	+7
1161	Feb. 22	Feb. 26	+4
1164	Mar. 16	Mar. 23	+7
1165	Mar. 30*		
1166	Mar. 25	Apr. 1	+7
1167	Mar. 6**		
1168	Apr. 2	Apr. 12	+10
1169	Mar. 1	Mar. 8	+7
1171	Mar. 19	Mar. 26	+7
1186	Mar. 8**		
1188	Mar. 20**		
1189	Mar. 1*		
1190	Mar. 19	Mar. 19	0
1191	Feb. 26	Mar. 15	+17
1193	Mar. 26	Apr. 2	+7
1197	Mar. 10*		
1199	Mar. 6	Mar. 13	+7
1200	Mar. 15	Mar. 22	+7
1204	Mar. 1**		
1207	Mar. 2	Mar. 9	+7
1208	Mar. 2	Mar. 9	+7
1211	Mar. 9	Mar. 16	+7
1213	Mar. 9	Mar. 18	+9
1216	Mar. 2	Mar. 9	+7
1217	Mar. 23	Mar. 30	+7
1223	Apr. 12*		
1224	Apr. 6	Mar. 14	−23
1225	May 9**		
1230	Mar. 6**		
1231	Mar. 17	Mar. 24	+7
1233	Apr. 18	May 16	+28
1234	Mar. 5	Mar. 12	+7
1235	Mar. 22	Mar. 29	+7
1238	Mar. 6*		
1246	Mar. 1	Mar. 8	+7
1253	Mar. 4	Mar. 11	+7
1254	Apr. 4	Apr. 11	+7
1259	Mar. 1	Mar. 8	+7
1264	Mar. 5	Mar. 12	+7

* from Song Hui Yao Ji Gao (Xu nineteenth century).
** from History of the Song Dynasty, section on five elements (Tuo, 1343).

in the year than those used previously. This revised result is very significant for the re-evaluation of the previous conclusion which indicated that a cold period occurred around A.D. 1200 (Chu, 1925, 1973). The previous conclusion is based on statistics of the latest snowfall date for both of the South Song period and the present. It was pointed out that "the average date of latest snowfall for each decade during the South Song time was April 9, almost a month later than the date of the latest spring snowfall for a decade in the first half of the twentieth century", and "Hangzhou was 1–2 °C colder in terms of the monthly mean temperature for April during the South Song time" (Chu, 1973). The re-examination of this record resulting from the revised calendar conversion indicates that the renewed date is, on average, eight days earlier than previously thought, and the earliest case is moved by 28 days. According to the revised data, the average date of the latest snowfall for each decade during the South Song time was 27 March, not April 9. Thus, the previous conclusion about estimated temperature for April in the South Song time should be revised. At least, the South Song time (around A.D. 1200) can not be considered as one of the coldest periods in the last 5000 years in China, as previously indicated. In this way, a contradiction with the nominal time interval of the Medieval Warm Period (A.D. 900–1300) may be partly resolved.

5. Conclusions

During the last 1300 years, the northern boundary of the cultivation of citrus trees and *Boehmeria nivea* has moved. The northernmost extent of cultivation of both plants occurred in the thirteenth century. As both plants are subtropical and sensitive to variations in thermal conditions, the thirteenth century can be considered as the warmest interval in this period. Temperature conditions in central China during the thirteenth century can be inferred on the basis of current climatic regionalization for citrus and *Boehmeria nivea*, and the thirteenth century locations of citrus crops and places where *Boehmeria nivea* yielded three crops per year. It is suggested that annual mean temperatures in the thirteenth century were 0.9–1.0 °C higher than present, monthly mean temperatures in January, and mean extreme minimum temperatures were 0.6 °C and 3.5 °C higher than present respectively. On the basis of the revised data for spring snowfall dates in Hangzhou (A.D. 1131–1264) it is pointed out that the previous conclusion that the South Song time was one of the coldest in recent millennia should be revised.

Acknowledgements

This research is supported by the National Science Foundation of China. I am indebted to Dr. Malcolm K. Hughes for his kind help in revising the manuscript in English. Thanks are also due to Dr. Marjorie G. Winkler for her help in preparing this paper.

References

Cheng Chunsu: 1991, *Climate and Agriculture in China*, Meteorology Press, Beijing, p. 275, (in Chinese).

Chen Yuan: 1962, *The Table of Intercalary Months and Date in the Twenty Histories*, Zhong-Hua Publishing House, Beijing, pp. 138–150.

Chinese Academy of Agriculture Science: 1980, 'A Study on Climatic Regionalization of Citrus in China', *Agrometeorol. J.* **2**, 13–18, (in Chinese).

Chu C. C.: 1925, *Collected Papers of Dr. Zhu K. Z.*, Science Press, Beijing, p. 53, (in Chinese).

Chu Co-Ching: 1931, 'Climatic Changes during Historical Times in China', *Gerlands Beitrage zur Geophysik, Koppen*, Band 1, 33, 29–27, Leipzig.

Chu Co-Chen: 1973, 'A Preliminary Study on the Climatic Fluctuations during the Last 5000 Years in China', *Sci. Sinica* **14**, 226, 236, 237.

Fang Shimin: 1987, *Conversion Table of the Chinese Historical Calendar*, Ci-Shu Publishing House, Shanghai, (in Chinese).

Grove, J.: 1988, *The Little Ice Age*, Methuen, London and New York.

Han Xiangling: 1991, *Crop Ecology*, Meteorology Press, Beijing, pp. 199–205.

He Kang: 1985, *National Atlas of Agriculture in China*, Photography Publishing House, Beijing, p. 114.

Ou Yangxiu: 1061, *New Tang Dynasty Book*, Section on geography, pp. 35–40 (in Chinese).

Tan qixiang: 1982, *The Historical Atlas of China*, Photography Publishing House, Beijing, pp. 5–7.

Tuo Tuo: 1343, *History of the Song Dynasty*, Section on geography 85–90, section on five elements 167–184 (in Chinese).

Xu Song: (Nineteenth Century), *Song Hui Yao Ji Gao*, Zhong- Hua Publishing House, Beijing, pp. 2089–2090, (in Chinese).

Wang Pan: 1273, *Nong Sang Ji Yao*, (Thread-bound Chinese book).

Zhang De'er: 1991, 'The Little Ice Age in China and Its Correlations with Global Change', *Quatern. Sci.* **3**, 104–106, (in Chinese).

Zhang De'er: 1992, 'The Little Ice Age in China', *CODATA Bull.* **4**, 91–100.

MAJOR WET INTERVAL IN WHITE MOUNTAINS MEDIEVAL WARM PERIOD EVIDENCED IN δ^{13}C OF BRISTLECONE PINE TREE RINGS

STEVEN W. LEAVITT

Laboratory of Tree-Ring Research, The University of Arizona, Tucson, AZ 85721, U.S.A.

Abstract. A long δ^{13}C chronology was developed from bristlecone pine (*Pinus longaeva*) at the Methuselah Walk site in the White Mountains of California. The chronology represents cellulose from five-year ring groups pooled from multiple radii of multiple trees. The most dramatic isotopic event in the chronology appears from A.D. 1080–1129, when δ^{13}C values are depressed to levels $\sim 2\sigma$ below the mean for the period A.D. 925–1654. This isotopic excursion appears to represent a real event and is not an artifact of sampling circumstances; in fact, a similar excursion occurs in a previously-reported, independent δ^{13}C chronology from bristlecone pine. By carbon isotope fractionation models, the shift to low δ^{13}C values is consistent with abundant soil moisture, permitting leaf stomata to remain open, and allowing ready access of CO_2 from which carbon fixation may discriminate more effectively against ^{13}C in favor of ^{12}C. According to this model, the ^{13}C-depleted 50-yr isotopic excursion represents the wettest period in the White Mountains in the past 1000 yr, during which isotope-reconstructed July Palmer Drought Severity Indices averaged $\sim +2.2$.

1. Introduction

Measurements of δ^{13}C (representing ^{13}C/^{12}C ratios) of tree rings have been made to reconstruct changes in δ^{13}C of atmospheric CO_2 (e.g., Stuiver, 1978; Freyer and Belacy, 1983; Stuiver *et al.*, 1984; Leavitt and Long, 1988) and to reconstruct climate (e.g., Ramesh *et al.*, 1986; Leavitt and Long, 1989a). The difficulties in reconstructing these parameters are related to complexity of the stable-carbon isotope fractionation models (Farquhar *et al.*, 1982) which indicate that, in addition to δ^{13}C of air CO_2, any factor influencing the ratio of intercellular CO_2 to atmospheric CO_2 concentrations can influence plant δ^{13}C. The ratio is driven by rates of carbon fixation and stomatal conductance and, therefore, is linked to factors such as light, moisture, relative humidity and CO_2 concentration. The drought-δ^{13}C link is possible through moisture deficits inducing decreased stomatal conductance, and hence, reduction of the concentration of CO_2 available for photosynthesis. A reduced intercellular CO_2 concentration would result in diminished isotopic discrimination against $^{13}CO_2$ by carbon-fixing enzyme, and consequently elevated values of plant δ^{13}C.

In this study, the inverse relationship of δ^{13}C values of recent bristlecone pine tree rings and drought indices (higher δ^{13}C $\Rightarrow$ more negative drought index values), is used to examine possible moisture anomalies during the Medieval Warm Period.

Climatic Change **26**: 299–307, 1994.
© 1994 *Kluwer Academic Publishers.*

The $\delta^{13}C$ values at A.D. 1100±25 represent an extraordinary negative departure from the mean, and suggest a very moist episode in the White Mountains within the Medieval Warm Period.

2. Field and Laboratory Methods

Bristlecone pine trees (*Pinus longaeva*) were cored in June 1984 in the White Mountains of east-central California (37°23'N, 118°09'W) at an elevation of ca. 3000 m (the 'Methuselah Walk' site). Nine trees were cored, 8 with 4 orthogonal cores, and the 9th, with two opposite cores. The rings of the cores were dated at the Laboratory of Tree-Ring Research; based on age stratification and ring characteristics, I chose four of the youngest trees to develop a $^{13}C/^{12}C$ chronology from A.D. 1420–1984 ('METH A'), and the 2 oldest, to develop an isotope chronology from A.D. 925–1654 ('METH B'). Ring limitations of these older trees prevented dating after the 1650's, but an overlap period of 235 yr allowed for assessing the potential for splicing the two chronologies. The rings were subdivided into 5-yr groups (...1895–1899, 1900–1904, 1905–1909,...) with a razor knife under a binocular microscope. The pentad samples were pooled from cores for METH A (4 cores from 4 trees) and METH B (4 cores from 2 trees) chronologies to produce site-representative isotopic chronologies (Leavitt and Long, 1984). After milling, oils and resins were leached from the samples with toluene-ethanol in a soxhlet extraction device, and cellulose was isolated by delignifying each sample in a 70 °C, acetic acid-acidified, sodium chlorite solution (after Green, 1963).

Samples were converted to CO_2 and H_2O at 800 °C in a microcombustion line in the presence of O_2 and CuO, and the CO_2 was cryogenically purified for mass-spectrometric analysis. By convention, the $\delta^{13}C$ * of each sample was calculated with respect to the PDB standard (Craig, 1957). During the period when these samples were analyzed, repeated combustion and analysis of a cellulose standard gave a precision of ±0.30‰ (±1 standard deviation). Separate processing and analysis of approximately every tenth pentad revealed a typical range of 1–2‰ among trees at the site. As reported in Leavitt and Long (1989b), however, the relative order of absolute $\delta^{13}C$ values among the trees generally remains the same regardless of pentad, and the trend in any one tree is similar to the mean trend of all trees.

$$ * \ \delta^{13}C \text{(in permil [‰] units)} = \left[\frac{^{13}C/^{12}C_{\text{sample}}}{^{13}C/^{12}C_{\text{PDBstandard}}} - 1 \right] \times 1000 $$

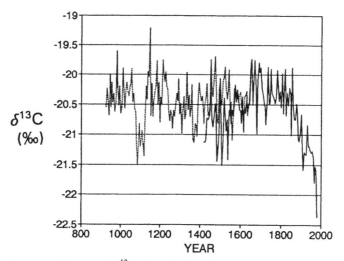

Fig. 1.   The bristlecone pine pentad $\delta^{13}C$ chronologies of METH A (solid) and METH B (dashed).

3. Results and Discussion

Both $\delta^{13}C$ chronologies are plotted in Figure 1. The two most notable aspects of the curves are the trend toward progressively more negative $\delta^{13}C$ values (^{13}C-depletion) after A.D. 1800 and the temporary excursion to more negative isotopic values during A.D. 1080–1129. The post-1800 decline is consistent with that found in many trees in the Americal Southwest and worldwide, and is interpreted as a signal of changing atmospheric $\delta^{13}C$ due to fossil-fuel inputs and shifts in land use (Leavitt and Long, 1988; Peng *et al.*, 1983). The excursion at A.D. 1100 is a $\delta^{13}C$ decline of ~2 σ relative to the mean of all METH B $\delta^{13}C$ values. One possible cause is that the pooling of cores, each with a different inside age and somewhat different absolute $\delta^{13}C$ values, may contribute to a $\delta^{13}C$ sampling artifact. This is probably the case in the overlap period of METH A and METH B (Figure 2), where the period 1550–1554 to 1650–1654 shows good correspondence in both absolute $\delta^{13}C$ (mean difference 0.04 ± 0.22‰) and fluctuations $r^2 = 0.37$; $P < 0.01$), and the period 1420–1424 to 1545–1549 shows good correspondence of fluctuations ($r^2 = 0.42$; $P < 0.01$) but not of absolute values (mean difference 0.54 ± 0.30‰). In this case, the number of trees contributing to the METH A pool changed from 3 trees after 1545–1549 to 2 trees before (all 4 trees of METH A are represented after 1655–1659). Because $\delta^{13}C$ of the tree that dropped out of the pool at A.D. 1545 was somewhat less negative than the remaining two, this likely contributed to the isotopic shift in METH A relative to METH B. However, at A.D. 1100 there is no change in pooling contribution that could similarly promote the abrupt displacement of $\delta^{13}C$ in the METH B chronology.

That this $\delta^{13}C$ excursion is representative, at least of the mid- to lower elevation bristlecone pine of the White Mountains, is confirmed in the 1800-yr $\delta^{13}C$ chronol-

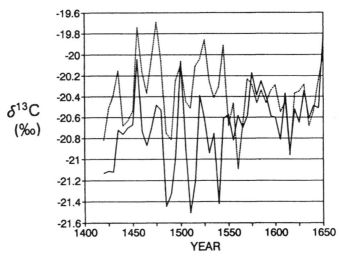

Fig. 2. The δ^{13}C chronologies of METH A (solid) and METH B (dashed) for the period over which they overlap, A.D. 1420–1654.

ogy of Stuiver *et al.* (1984). Their δ^{13}C chronology also shows a dramatic 50-yr excursion centered at A.D. 1100, yet it was developed from cellulose of decadal tree-ring groups from one radius of a single tree ∼ 7 km north of the Methuselah site.

The high-frequency δ^{13}C fluctuations in the METH chronologies are dominated by moisture influences. This conclusion follows from comparison of the δ^{13}C with climatological measures of drought (Palmer, 1965; Karl *et al.*, 1983) such as the Palmer Hydrologic Drought Indices [PHDI] and Palmer Drought Severity Indices [PDSI]. These indices are calculated from meteorological parameters and are measures of moisture abnormality. The indices generally fall between values of +6 to -6, representing gradation from extreme wet conditions to extreme drought, respectively, with an index of 0 representing 'normal' moisture conditions. PDSI is commonly considered to be a meteorological drought index, whereas PHDI is more of a hydrological drought index. PDSI has been used more frequently in climate reconstructions using tree-ring widths.

Pentad (60-month) averages of PHDI from 1931–1982 of the Southeast Desert Basins climatic subdivision, in which the White Mountains are located, had already been regressed against δ^{13}C measures in Leavitt and Long (1989a). Rather than using δ^{13}C directly, because of the declining δ^{13}C trend during this comparison period, a curve (spline) was fit to the METH A δ^{13}C data and an index value (Del Index) was calculated as a function of the measured and smoothing curve values:

$$\text{Del Index} = \left[\frac{\delta^{13}C_{\text{measured}}}{\delta^{13}C_{\text{smoothed curved}}} - 1 \right] \times 1000 \tag{1}$$

This index provides more comparable large numbers (as opposed to decimal

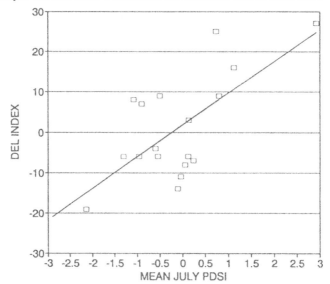

Fig. 3.   Relationship of METH A Del Indices with corresponding average July PDSI (Northeast Interior Basins, CA) for the pentads from A.D. 1895–1983. The slope = 7.8 and $r^2 = 0.46$ $P < 0.01$).

fractions from the simple ratio), with negative values for moisture 'deficiency' and positive values for moisture 'excess' according to theory. This model's success was evidenced by high correlations of Del Indices of pinyon pine from around the Southwest with respective climate subdivision PHDI, but the correlation coefficient for METH A Del Indices with PHDI (1931–1982) of its Southeast Desert Basins (California) subdivision was actually not significant ($r^2 = 0.01$) (Leavitt and Long, 1989). Comparison of METH A Del Indices with pentad-averaged July PHDI (1931–1982) from three proximate subdivisions (Southeast Desert Basins California, South Central Nevada and Northeast Interior Basins California) yields $r^2 = 0.00$ (ns), 0.41 ($P < 0.05$) and 0.68 ($P < 0.01$), respectively. This result suggests that despite residing in the Southeast Desert Basins California subdivision, the White Mountains are at the extreme northern edge of a subdivision which extends south to Mexico, and the White Mountains climate might be better approximated by that in the other two subdivisions, especially the Northeast Interior Basins. Because of the short growing season in the White Mountains (Fritts, 1976), July drought indices are probably a better measure for comparison than the full-year averages. Employing July PDSI records back to 1895, although potentially of poorer quality than the post-1930 PDSI, correlation coefficients of METH A Del Indices with Southeast Desert Basins, South Central Nevada and Northeast Interior Basins are $r^2 = 0.13$ (ns), 0.33 ($P < 0.05$) and 0.46 ($P < 0.01$), respectively. Figure 3 depicts the relationship of METH A Del Indices with Northeast Interior Basins July PDSI for the 1895–1983 period.

   Tree-ring width time series can often be modeled with a negative exponential growth curve for southwestern trees. There are no similar expectations for $\delta^{13}C$

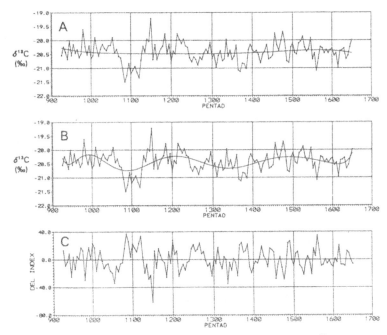

Fig. 4. Fifth-degree (A) and 10th-degree (B) polynomial fits to METH B δ^{13}C values, and the Del Indices (C) calculated from the 10th-degree polynomial curve with Equation (1).

values in tree rings during their lifetime. Thus, spline and polynomial curves seem to be legitimate tools for modelling the low-frequency δ^{13}C changes, especially the post-1800 δ^{13}C decline, in order to isolate the high-frequency climate effects. Del Indices were developed for the METH B chronology by fitting a polynomial curve and using the algorithm in Equation (1) (Figure 4). Even using a more flexible 10th-degree polynomial (Figure 4B) which enhances medium frequency trends in the δ^{13}C chronology, the Del Indices (Figure 4C) show the period, A.D. 1080–1129, as anomalously moist in the record, although it is flanked by two very dry periods. Del Indices calculated from a less flexible 5th-degree polynomial (Figure 4A) would not be much different than those from a straight-line regression and they would make the 1080–1129 period seem even more anomalously wet than adjacent periods before and after, in terms of Del Indices.

Figure 5 shows a plot of ring-width indices from a site ~ 4 km north and ~ 250–300 m higher (White Mountains chronology). The apparently large year-to-year ring-width index variation is just what one expects for a 'sensitive' tree-ring series, and enhances its potential for climate reconstruction from ring size (Fritts, 1976). Within A.D. 900–1300, a period of substantially enhanced growth is centered around A.D. 1100 characterized by index values greater than the average growth index of 1; this is especially evident in the 11-yr running mean. The Methuselah Walk chronology (D. Graybill, 1991, pers. comm.), virtually at the site of the isotopic sample collection, also shows a period of greater-than-normal growth

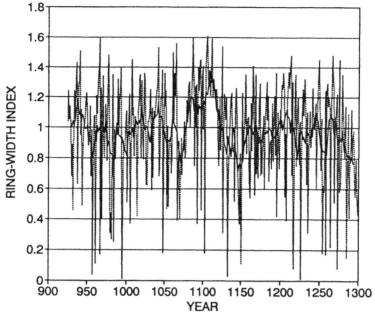

Fig. 5. The 'White Mountains' bristelcocne pine tree-ring width index chronology (dashed) (Drew, 1972) with an 11-yr running mean fit to the data (solid). Values above '1' indicate above-normal growth, and those below '1' are below normal.

(ring indices > 1) at A.D. 1100 relative to the several hundred years before and after.

Stuiver *et al.* (1984) presented a $\delta^{13}C$ chronology (pentads) from a single giant sequoia tree from the neighboring Sierra Nevada Mountains to the west. It records no excursion in $\delta^{13}C$ around A.D. 1100. Further, the whole period, A.D. 800–1300, is characterized by a $\delta^{13}C$ maxima in the long-term trend, with one major short-term departure to lower $\delta^{13}C$ values at ~ A.D. 1175. However, major climatic differences exist between the White Mountains and the Sierra Nevada; precipitation of the latter more dominated by winter storms. Another proxy record, that of Mono Lake levels developed by Stine (1990) does show a sudden and substantial rise in lake level at ~ A.D. 1100 based on ^{14}C dates on tree stumps that apparently were drowned during a short period of abnormally high streamflow into the lake. A problem arises in precise calendric dating of the trees because of high uncertainties associated with calibration of the ^{14}C time scale at this time period, and because they have not been datable by dendrochronology.

Finally, Epstein and Yapp (1976) published a δD chronology from bristlecone pine (decadal samples) for the period, A.D. 970–1974. The elevation of the chronology was given as 3000 m, which corresponds approximately to the site described herein. The most notable feature of their curve is isotopically light (more negative) δD values for A.D. 1450–1650, which they interpreted as a Little Ice Age signal. A short departure to less negative δD values centers around A.D. 1100. The standard interpretation of δD in plant cellulose is that less negative values imply warmer conditions and more negative values imply cooler conditions, although there are currently some questions as to the exact interpretation of δD values (DeNiro and Cooper, 1989). If the δD curve does represent temperature, then, in conjunction with the $\delta^{13}C$ chronology, the implication is a short (50-yr) warmer and wetter period in the White Mountains around A.D. 1100.

3.1. *Conclusions*

The most notable feature of the thousand-year $\delta^{13}C$ chronology from bristlecone pine, besides a post-A.D. 1800 decline, is the major negative excursion between A.D. 1080 and 1129. An atmospheric CO_2 $\delta^{13}C$ excursion of this magnitude and short length would seem unlikely, especially since a coincident decline is not evident in the sequoia of Stuiver *et al.* (1984). Ring-width indices indicate several years of very low growth (low precipitation) in this 50-yr period, but overall the period was characterized by much greater growth than the several hundred years before and after. Strong positive correlations of average July PHDIs and PDSIs with Del Indices derived from detrending the $\delta^{13}C$ values suggest A.D. 1080-1129 was characterized by extremely abundant soil moisture. Using the regression equation between Northeast Interior Basin July PDSI and METH Del Indices, the average Del Indices of +19.3 in this period translate to average July PDSI of $\sim$ +2.2 for these 50 yr. This isotopic response is consistent with the expected effects of soil moisture availability on stomatal conductance and its control on CO_2 entry. Thus, in the White Mountains of California, the Medieval Warm Period contained a brief but pronounced wet excursion (and perhaps somewhat warmer as evidenced by the hydrogen isotopes) which appears to fall between two periods of very dry conditions of similar duration. The ring-width index patterns do not reveal the full magnitude of this wet event evidenced in the $\delta^{13}C$ chronology, perhaps because once a maximum ring size is attained under favorable growth conditions, any further improvement in conditions (e.g., more water) cannot increase growth further. This suggests that, in spite of such limitations on maximum ring size, the isotopes may continue to be a valuable proxy indicator of physiological responses to moisture conditions when ring size no longer responds to such extreme climatic conditions.

Acknowledgements

Austin Long, Songlin Cheng, Lisa Warneke and Bob Kalin helped in collecting the samples and/or developing the isotope chronologies. Denny Bowden dated the tree rings, and Don Graybill provided information from the most recent version of the Methuselah bristlecone pine ring-width index chronology. Development of the bristlecone $\delta^{13}C$ chronologies was supported by Oak Ridge/Martin Marietta subcontract no. 19X-22290C through the U.S. Dept. of Energy Carbon Cycle Research Program.

References

Craig, H.: 1957, 'Isotopic Standards for Carbon and Oxygen and Correction Factors for Mass-Spectrometric Analysis of CO_2', *Geochim. Cosmochim. Acta* **12**, 133–149.

DeNiro, M. J. and Cooper, L. W.: 1989, 'Post-Photosynthetic Modification of Oxygen Isotope Ratios of Carbohydrates of Potato: Implications for Paleoclimate Reconstruction Based on Isotopic Analysis of Wood Cellulose', *Geochem. Cosmochim. Acta* **53**, 2573–2580.

Drew, L. G. (ed.): 1972, *Tree-Ring Chronologies of Western America III California and Nevada*, Lab. of Tree-Ring Research, University of Arizona, Tucson, Chronology Series 1.

Epstein, S. and Yapp, C. J.: 1976, 'Climatic Implications of the D/H Ratio of Hydrogen in C-H Groups in Tree Cellulose', *Earth Planet Sci. Lett.* **30**, 252–261.

Farquhar, G. D., O'Leary, M. H., and Berry, J. A.: 1982, 'On the Relationship between Carbon Isotope Discrimination and the Intercellular Carbon Dioxide Concentration in Leaves', *Austr. J. Plant Phys.* **9**, 121–137.

Freyer, H. D. and Belacy, N.: 1983, '$^{13}C/^{12}C$ Records in Northern Hemispheric Trees during the Past 500 Years. Anthropogenic Impact and Climate Superpositions', *J. Geophys. Res.* **88**, 6844–6852.

Fritts, H.: 1976, *Tree Rings and Climate*, Academic, New York, 567 pp.

Green, J. W.: 1963, 'Wood Cellulose' in Whistler, R. L. (ed.), *Methods of Carbohydrate Chemistry*, Academic, New York, pp. 9–21.

Karl, T. R., Metcalf, L. K., Nicodemus, M. L., and Quayle, R. G.: 1983, *Historical Climatology Series 6–1: Statewide Average Climatic History*, NOAA, National Climate Data Center, Asheville, North Carolina.

Leavitt, S. W. and Long, A.: 1984, 'Sampling Strategy for Stable Carbon Isotope Analysis in Pine', *Nature* **311**, 145–147.

Leavitt, S. W. and Long, A.: 1988, 'Stable Carbon Isotope Chronologies from Trees in the South-western United States', *Glob. Biogeochem. Cycl.* **2**, 189–198.

Leavitt, S. W. and Long, A.: 1989a, 'Drought Indicated in Carbon-13/Carbon-12 Ratios of South-western Tree Rings', *Water Res. Bull.* **25**, 341–347.

Leavitt, S. W. and Long, A.: 1989b, 'Intertree Variability of $\delta^{13}C$ in Tree Rings', in Rundel, P. W., Ehleringer, J. R., and Nagy, K. A., (eds.), *Stable Isotopes in Ecological Research*. Springer-Verlag, New York, pp. 95–104.

Palmer, W. C.: 1965, *Meteorological Drought*, U.S. Weather Bureau Research Paper No. 45. U.S. Dept. of Commerce, Washington, D.C.

Peng, T. H., Broecker, W. S., Freyer, H. D., and Trumbore, S.: 1983, 'A Deconvolution of the Tree-Ring Based $\delta^{13}C$ Record', *J. Geophys. Res.* **88**, 3609–3620.

Ramesh, R., Bhattacharya, S. K., and Gopalan, K.: 1986, 'Climatic Correlations in Stable Isotope Records of Silver Fir (*Abies pindrow*) Trees from Kashmir, India', *Earth Planet. Sci. Lett.* **79**, 66–74.

Stine, S.: 1990, 'Late Holocene Fluctuations of Mono Lake, Eastern California', *Palaeogeog. Palaeoclimat. Palaeoecol.* **78**, 333–381.

Stuiver, M.: 1978, 'Atmospheric Carbon Dioxide and Carbon Reservoir Changes', *Science* **199**, 253–258.

Stuiver, M., Burk, R. L., and Quay, P. D.: 1984, '$^{13}C/^{12}C$ Ratios in Tree Rings and the Transfer of Biospheric Carbon to the Atmosphere', *J. Geophys. Res.* **89**, 11731–11748.

(Received 22 September, 1992; in revised form 11 October, 1993).

THE MEDIEVAL SOLAR ACTIVITY MAXIMUM

J. L. JIRIKOWIC and P. E. DAMON

Laboratory of Isotope Geochemistry and the NSF-Arizona Accelerator Facility for Isotope Analysis, Department of Geosciences, University of Arizona, Tucson, Arizona 85721, U.S.A.

Abstract. Paleoclimatic studies of the Medieval Solar Maximum (*c.* A.D. 1100–1250, corresponding with the span of the Medieval Warm Epoch) may prove useful because it provides a better analog to the present solar forcing than the intervening era. The Medieval Solar Activity Maximum caused the cosmogenic isotope production minimum during the 12th and 13th Centuries A.D. reflected by Δ^{14}C and ^{10}Be records stored in natural archives. These records suggest solar activity has returned to Medieval Solar Maximum highs after a prolonged period of reduced solar activity. Climate forcing by increased solar activity may explain some of this century's temperature rise without assuming unacceptably high climate sensitivity. By analogy with the Medieval Solar Activity Maximum, the contemporary solar activity maximum may be projected to last for 150 years. The maximum temperature increase forced by increased solar activity stays well below the predicted doubled atmospheric CO_2 greenhouse forcing.

1. Introduction

While summarizing the *Medieval Warm Period Workshop*, Dr. Malcolm Hughes stated the most definitive description of the Medieval Warm Period available at this time, "... at least, it wasn't a Little Ice Age" (Hughes, personal communication, 1991) Dr. Hughes succinctly expressed the one conclusion that most participants appeared to reach during the workshop. This resolution logically defines the paleo-climatic discussion of a Medieval Warm Period variation as a contrast with climate since the 13th Century A.D.

Although solar activity variations have been noted since the inception of astronomy, Eddy's (1976) discussion of Maunder's historical evidence for the 17th Century Solar Activity Minimum (Maunder Minimum) revived the search for similar solar activity minima in historical and natural records. The sharp contrast between the Maunder Minimum and the modern state of solar activity provides the most interesting conclusion of Eddy's (1976) work. Since Eddy's paper, work exploring natural cosmogenic isotope archives proxying for solar activity data, primarily ^{14}C and ^{10}Be, have largely focused upon the isotope activity maxima that correspond to solar activity minima (see Damon and Sonett, 1991, for review).

From these cosmogenic isotope-solar activity archives, one can define a high solar activity interval distinct from the Maunder or any other solar activity minimum termed the Medieval Solar Maximum (A.D. 1100–1250, Damon and Jirikowic, 1992a). Moreover, after 650 years of reduced solar activity, solar activity has returned to Medieval Solar Maximum levels during this century. Thus, the Medieval

Climatic Change **26**: 309–316, 1994.
© 1994 *Kluwer Academic Publishers.*

TABLE I: Parameters for our solar activity model of global equilibrium surface temperature forcing after Damon and Jirikowic (1992b). ACRIM observations of Willson and Hudson (1988) constrain the 11-year period solar irradiance variation. Matching the 20th Century temperature increase fixes the 88- and 208-year period amplitudes. Damping factors and lag times from a climate model by Wigley (1988)

Period (yrs)	$\%S_0$	ΔS (Wm^{-2})	Damping Factor	Lag (yrs)	$\Delta\theta eq$ (°C)
11	±0.04	±0.55	0.24	1.6	±0.010
88	±0.23	±3.10	0.61	7.5	±0.150
208	±0.37	±5.11	0.74	13.2	±0.300

Solar Maximum provides a solar-terrestrial analog for investigating the constraints of solar forcing on an un-anthropogenically disturbed climate.

The upper limit of solar-forced global average temperature increase since the beginning of this century has been estimated to be 0.6 °C (Damon and Jirikowic, 1992b). A model based upon the periodicities with their respective phases present in the tree-ring Δ^{14}C record (Damon and Sonett, 1991) predicts a solar activity maximum beginning near A.D. 1900 ending near A.D. 2050. The maximum contribution to total global warming would be less than 0.8 °C (see Figure 1; Table I; Damon and Jirikowic, 1992b). Mechanisms for solar forcing of climate include total luminosity, spectral luminosity and particle radiation variations (see Schatten and Arkin, 1990, for review). The pre-industrial (*ca.* A.D. 1850) similarity between the ^{10}Be and Δ^{14}C signatures suggests that similar solar-driven interplanetary conditions existed during the present Contemporary Solar Maximum and the Medieval Solar Maximum (see Figure 2).

The Medieval Solar Maximum roughly coincides with the Medieval Warm Epoch (A.D. 1000–1200; after Lamb, 1965). The Medieval Warm Epoch provides a climate scenario to the 20th Century A.D. without anthropogenic perturbations. Our study of the Medieval Solar Maximum will elaborate upon these three points:

— the Medieval Solar Maximum provides an apt solar-terrestrial analog to the present climate had it not been disturbed by anthropogenic forcings;

— solar activity during the Medieval Solar Maximum reached modern levels and remained at those levels for a total duration of about 150 years, and;

— increased solar activity forcing of climate may cause an increasing global surface temperature ≈0.8 °C at most, well below the projected ≈1.5–5.5 °C increase for doubled atmospheric CO_2 concentration (from Hansen and Lacis, 1990).

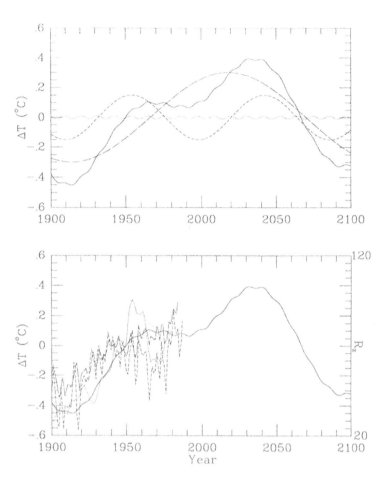

Fig. 1. The upper panel shows a model curve developed by assigning forcings of 0.1, 0.5 and 0.9 $W \cdot m^{-2}$ to the 11- (Schwabe), 88- (Gleissberg), and 208-year (Suess) solar cycles (dotted, short-dash and long-dash, respectively) and from the parameters in Table I. The amplitudes of the 88- and 208-year forcings were chosen to fit most closely the 20th Century temperature trend. The forcings linearly combined into the solid line curve to model the upper limit of bolometric variation forced equilibrium surface temperature change in the 20th and 21st Centuries. The maximum solar activity forcings of 1.5 $W \cdot m^{-2}$ is smaller than the 2 $W \cdot m^{-2}$ estimated greenhouse forcing to 1989 and much smaller than doubled greenhouse gas forcing of 4–4.8 $W \cdot m^{-2}$. The lower panel compares estimates of 20th Century global temperature change. The model curve remains within the envelop of annual temperature variability expressed in Jones *et al.* (1986, long-dash) and Hansen and Lebedeff (1988, short-dash) data. The model curve generally lags temperature, however. Note that the envelope of the 11-year running average of the Wolf sunspot indices, R_z (dotted), follow both our model and the global temperature change (Reid and Gage, 1988). Projection of our model curve into the 21st Century shows that the total model temperature increase from the beginning of the 20th Century will remain below 0.8 °C, well below estimates of greenhouse warming. The model predicts this contemporary high solar activity interval will last about 150 years, similar to the duration of the Medieval Warm Period (A.D. 1100–1250).

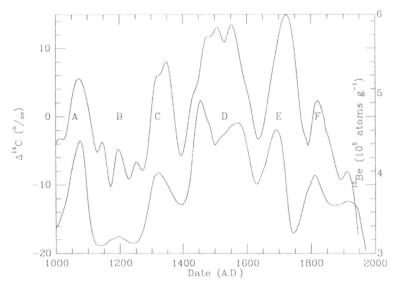

Fig. 2.   Tree-ring $\Delta^{14}C$ (upper line, from Stuiver and Pearson, 1986) and South Polar ice ^{10}Be concentration variations (lower line, from Raisbeck *et al.*, 1990) during the last millennium. The letter **B** marks the Medieval Solar Maximum, **A, C, D, E** and **F** mark the Oort, Wolf, Spörer, Maunder and Dalton minima, respectively. The curves represent splines (stiffness = 5) through the data. Note the close correspondence between the two curves. Since for these geochemically disparate isotopes both data sets show similar variations and very limited differential response to regional climatic fluctuations, the century-scale variations are considered to have resulted from varying atmospheric cosmogenic isotope production. High solar activity is accompanied by strong solar wind coupling with the geomagnetic field that shields the atmosphere from cosmogenic isotope-producing galactic cosmic rays. Thus a cosmogenic isotope minimum occurs during the Medieval Solar Maximum. Also note that 20th Century ^{10}Be concentrations have returned to their Medieval Solar Maximum nadir suggesting the contemporary solar maximum will be much as the Medieval Solar Maximum.

2. Cosmogenic Isotope Indications of the Medieval Solar Maximum

Cosmogenic isotopic production varies inversely with the strength of the solar wind-geomagnetic field coupling. Magnetic fields deflect some electrically charged galactic cosmic ray particles before they enter the Earth's atmosphere. Those galactic cosmic ray particles that do reach the atmosphere initiate nuclear reactions that produce cosmogenic isotopes (see O'Brien *et al.*, 1991 for review). As the solar wind coupled-geomagnetic field strengthens, fewer galactic cosmic rays reach the atmosphere to produce cosmogenic isotopes. Thus, the Medieval Solar Maximum, with its high solar activity and consequently high solar wind magnetic field, resulted in a decrease in production of cosmogenic isotopes. Conversely, the Maunder Minimum, with its low solar wind magnetic field strength, caused an increase in the production of cosmogenic isotopes.

Two cosmogenic isotope archives have provided the best records of solar activity to date: ^{14}C in tree-rings and ^{10}Be in polar ice cores. ^{14}C-tagged atmospheric CO_2

taken up by photosynthesis and incorporated into the tree-ring cellulose not only provides an annual record of atmospheric ^{14}C activity, but also provides independent dendrochronologic dating. After corrections for radioactive decay and isotopic fractionation, Δ^{14}C reflects variations in atmospheric ^{14}C activity relative to a mid-19th Century reference.

[10]Be precipitates quickly out of the atmosphere, often within hygroscopic condensation nuclei. Provided ablation does not occur, ice cores record the [10]Be activity in snowfall. Single cyclonic storms may have a disproportionate influence upon the [10]Be activity in an annual ice layer. Cores within the climatically-stable Antarctic polar desert provide the most stable [10]Be archives as verified by the climatically-indicative δ^2H isotopic ratio (Beer *et al.*, 1991).

Figure 2 compares natural cubic spline curves through the decadal Δ^{14}C data (Stuiver and Pearson, 1986) and the South Pole [10]Be concentrations of Raisbeck *et al.* (1990). The correspondence between the century-scale features (annotated with letters) suggests these features record global atmospheric cosmogenic isotope activity variations that could only be caused by a similarly varying production rate. The Maunder Minimum with low solar activity resulting in weak solar wind magnetic fields inferred from sunspot and aurora observations, can be observed as a maximum in both cosmogenic isotope records (annotation **E**). The Medieval Solar Maximum, when high solar activity reduced the atmospheric production of cosmogenic isotopes, spans from A.D. 1100 to A.D. 1250 (annotation **B**). Immediately after A.D. 1250 and persisting for 650 years, higher Δ^{14}C suggests elevated atmospheric cosmogenic isotope production during much of this millenium. Such events recur in the Δ^{14}C archive and may show prolonged low-solar activity states (see Hood and Jirikowic, 1990 for example, Damon and Sonett, 1991 for review).

[10]Be concentrations also show this 650-year interval of high cosmogenic isotope production. Since the beginning of the 20th Century, [10]Be concentrations have returned to low concentrations similar to those during the Medieval Solar Maximum. The sharp decrease in Δ^{14}C during the 20th Century has been distorted by the dumping of ^{14}C-depleted carbon into the atmosphere by fossil fuel combustion. However, [10]Be concentrations provide clear evidence that solar activity has returned to Medieval Solar Maximum levels (anthropogenic production of [10]Be being almost nonexistent). If the Medieval Solar Maximum provides an analog to the present, the higher solar activities during this century can be expected to persist for ≈ 150 years.

3. The Medieval Solar Maximum as a Paleoclimatic Analog of a Pristine Contemporary Climate

Episodes such as montane glacier advances and retreats are synchronous between the Northern and Southern Hemispheres (Rothlisberger, 1986; Grove, 1988) suggests the global extent of the cooling and warming during climate episodes such as the Little Ice Age and Medieval Warm Period. A statistically significant correlation has also been found between two earlier episodes of glacial advance and retreat (Wigley and Kelly, 1990). Evidence shows recent glacial retreat surpasses these earlier events (Hastenrath and Kruss, 1992). The global consistency of these glacial retreats requires that the global climate forcings were of sufficiently long duration to overcome the considerable thermal inertia of montane glaciers. Along with anthropogenic forcings, prolonged variations in solar forcing meet this criterion.

The Medieval Solar Maximum represents the only pre-20th Century prolonged solar maximum during this millenium. From the end of the Medieval Solar Maximum until possibly as late of as the close of the 19th Century, solar activity remained profoundly depressed. In contrast, the 20th Century witnessed higher solar activity, much more analogous to the solar-terrestrial environment of the Medieval Solar Maximum.

As an exploration of the global average temperature response to the Medieval Solar Maximum, we begin with the models developed in an earlier study of 20th Century solar forcing of climate (see Figure 1; Table I; Damon and Jirikowic, 1992b). Using consecutive linear regression between cosmogenic isotope production and solar activity (after Stuiver and Quay, 1980), linear regression between solar activity and total solar irradiance (Willson and Hudson, 1988) and sensitivities, feedback and lag terms from Damon and Jirikowic (1992b), we estimated the maximum global temperature forced by total solar irradiance variations during the Medieval Solar Maximum. Not surprisingly, the total range is limited to below 0.8 °C (see Figure 3). Lean *et al.* (1992) arrived at a somewhat smaller upper bound for solar forcing using the comparison between contemporary solar activity phenomena and Maunder Solar Minimum scenarios.

4. Conclusions

As discussed by our co-participants in this volume and during the Medieval Warm Period Workshop, the 0.6 °C global warming since the beginning of the 20th Century may be unprecedented in extent and rapidity. We have shown that such a rise requires the largest possible assumed global temperature response to solar forcing. During the Holocene, global temperatures have apparently not varied so greatly and so quickly. Thus, although solar forcing may significantly contribute to Holocene century-scale climate variations, the anthropogenically-perturbed 20th Century clearly represents a change from the past pattern. Thus, warming during the Medieval Solar Maximum should not be considered a complete paleoclimatic analog to the present but as a baseline for climate variability without greenhouse forcing.

If greenhouse gas emissions are a geophysical experiment on a grand scale (albeit poorly designed), the Medieval Solar Maximum does provide a control run. This study of solar activity forcing suggests that Greenhouse forcing superimposed upon the present high-solar activity epoch dominates the global surface temperature increase during this century. Studies of the Medieval Warm Period paleoclimate

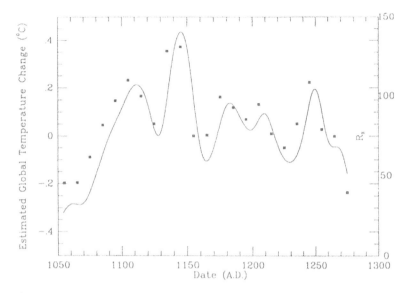

Fig. 3. Δ^{14}C-derived R_z estimate (solid squares) and global temperature model predictions (curve) based upon linear regressions of Stuiver and Quay (1980) and Willson and Hudson (1988) and the model parameters in Table I. The maximum global temperature rise remains below 0.8 °C and anomalously high temperatures persist for about 150 years, similar to predictions for the Contemporary Solar Maximum.

should suggest how the present climate would vary due to natural variations without anthropogenic forcing.

Acknowledgements

The authors benefited from discussions with Professors Charles P. Sonett and Lonnie L. Hood of the University of Arizona. This work was supported by NSF Grants ATM-9012102, ATM-8919535, EAR-8822292 and the State of Arizona.

References

Beer, J., Raisbeck, G. M., and Yiou, F.: 1991, 'Time Variations of ^{10}Be and Solar Activity', in Sonett, C. P., Giampapa, M. S., and Matthews, M. S. (eds.), *The Sun in Time*, University of Arizona Press, Tucson, 343 pp.

Bradley, R. S.: 1985, *Quarternary Paleoclimatology: Methods of Paleoclimatic Reconstruction*, Allen and Unwin, Boston, p. 472.

Damon, P. E. and Jirikowic, J. L.: 1992a, 'Radiocarbon Evidence for Low Frequency Solar Oscillations', in Povinec, P. (ed.), *Rare Nuclear Decay Processes: Proceedings of the 14th European Physics Society Meeting, Bratislava, October 1990*, 457 pp.

Damon, P. E. and Jirikowic, J. L.: 1992b, 'Solar Forcing of Global Climate Change?', In Taylor, R. E., Long, A., and Kra, R. (eds.), *Four Decades of Radiocarbon*, Springer-Verlag, New York, 117 pp.

Damon, P. E. and Sonett, C. P.: 1991, 'Solar and Terrestrial Components of the Atmospheric ^{14}C Variation Spectrum', in Sonett, C. P., Giampapa, M. S., and Matthews, M. S. (eds.), *The Sun in Time*, University of Arizona press, Tucson, 360 pp.

Eddy, J. A.: 1976, 'The Maunder Minimum', *Science* **192**, 1189–1202.

Eddy, J. A.: 1977, 'Climate and the Changing Sun', *Climatic Change* **1**, 173.

Grove, J. M.: 1991, *The Little Ice Age*, Routledge, London, p. 498.

Hansen, J. E. and Lacis, A. A.: 1990, 'Sun and Dust versus Greenhouse Gases: As Assessment of Their Relative Roles in Global Climate Change', *Nature* **346**, 713.

Hansen, J. E. and Lebedeff, S.: 1988, 'Global Surface Air Surface Temperatures Update through 1987', *Geophys. Res. Let.* **15**, 323.

Hastenrath, S. and Kruss, P. D.: 1992, 'Greenhouse Indicators in Kenya', *Science* **355**, 503.

Hood, L. L. and Jirikowic, J. L.: 1990, 'Recurring Variations of a Probable Solar Origin on the Atmospheric Δ^{14}C Record', *Geophys. Res. Let.* **17**, 85.

Jones, P. D., Wigley, T. M. L., and Wright, P. B.: 1986, 'Global Temperature Variation between 1862 and 1984', *Nature* **322**, 430.

Lamb, H. H.: 1965, 'The Early Medieval Warm Epoch and Its Sequel', *Paleogeog., Paleoclim., Paleoecol.* **1**, 13.

Lean, J., Skumanich, A., White, O.: 1992, 'Estimating the Sun's Radiative Output during the Maunder Minimum', *Geophys. Res. Let.* **19**, 1591.

O'Brien, K., de al Zerda Lerner, A., Shea, M. A., and Smart, D. F.: 1991, 'The Production of Cosmogenic Isotopes in the Earth's Atmosphere and Their Inventories', in Sonett, C. P., Giampapa, M. S., and Matthews, M. S. (eds.), *The Sun in Time*, Tucson, University of Arizona Press, 317 pp.

Raisbeck, G. M., Yiou, F., Jouzel, J., and Petit, J. R.: 1990, '^{10}Be and δ^2H in Polar Ice Cores as a Probe of the Solar Variability's Influence on Climate', *Phil. Trans. R. Soc. Lond.* **A330**, 463.

Reid, G. C. and Gage, K. S.: 1988, 'The Climatic Impact of Secular Variations in Solar Irradiance', in Stephenson, F. R. and Wolfdale, A. W. (eds.), *Secular, Solar, and Geomagnetic Variations in the Last 10,000 Years*, Kluwer Academic Press, Dordrecht, 225 pp.

Rothlisberger, F.: 1986, *10000 Jahre Gletschergeschichte der Erde*, Aarau, Verlag Sauerlander, p. 416.

Schatten, K. H. and Arking, A. (eds.): 1990, *Climate Impact of Solar Variability*, NASA Conference Publication 3086, p. 376.

Stuiver, M. and Pearson, G. W.: 1986, 'High-Precision Calibration of the Radiocarbon Time Scale A.D. 1950–500 BC', in Stuiver, M. and Kra, R. S. (eds.), *Radiocarbon* **28(2B)**, 805.

Stuiver, M. and Quay, P. D.: 1980, 'Changes in Atmospheric Carbon-14 Attributed to a Variable Sun', *Science* **207**, 11.

Wigley, T. M. L.: 1988, 'The Climate of the Past 10,000 Years and the Role of the Sun', in Stephenson, F. R. and Wolfdale, A. W. (eds.), *Secular, Solar, and Geomagnetic Variations in the Last 10,000 Years*, Kluwer Academic Press, Dordrecht, 209 pp.

Wigley, T. M. L. and Kelly, P. M.: 1990, 'Holocene Climate Change, ^{14}C Wiggles and Variations in Solar Irradiance', *Phil. Trans. R. Soc. of Lond.* **A330**, 547.

Wigley, T. M. L. and Raper, S. C. B.: 1987, 'Thermal Expansion of Sea Water Associated with Global Warming', *Nature* **330**, 127.

Willson, R. C. and Hudson, H. S.: 1988, 'Solar Luminosity Variations in the Solar Cycle', *Nature* **332**, 810.

(Received 22 September, 1993; in revised form 5 November, 1993)

AN ANALYSIS OF THE TIME SCALES OF VARIABILITY IN CENTURIES-LONG ENSO-SENSITIVE RECORDS IN THE LAST 1000 YEARS

HENRY F. DIAZ and ROGER S. PULWARTY

NOAA/ERL/CDC, 325 Broadway, Boulder, CO 80303, U.S.A., Cooperative Institute for Research in Environmental Sciences, and Department of Geography, University of Colorado, Boulder, CO 80309, U.S.A.

Abstract. We document the characteristic time scales of variability for seven climate indices whose time-dependent behavior is sensitive to some aspect of the El Niño/Southern Oscillation (ENSO). The ENSO sensitivity arises from the location of these long-term records on the periphery of the Indian and Pacific Oceans. Three of the indices are derived principally from historical sources, three others consist of tree-ring reconstructions (one of summer temperature, and the other two of winter rainfall), and one is an annual record of oxygen isotopic composition for a high-elevation glacier in Peru. Five of the seven indices sample at least portions of the Medieval Warm Period ($\sim$ A.D. 950 to 1250).

Time series spectral analysis was used to identify the major time scales of variability among the different indices. We focus on two principal time scales: a high frequency band ($\sim$ 2–10 yr), which comprises most of the variability found in the modern record of ENSO activity, and a low frequency band to highlight variations on decadal to century time scales ($11 < P < 150$ yr). This last spectral band contains variability on time scales that are of general interest with respect to possible changes in large-scale air-sea exchanges. A technique called evolutive spectral analysis (ESA) is used to ascertain how stable each spectral peak is in time. Coherence and phase spectra are also calculated among the different indices over each full common period, and following a 91-yr window through time to examine whether the relationships change.

In general, spectral power on time scales of $\sim$ 2–6 yr is statistically significant and persists throughout most of the time intervals sampled by the different indices. Assuming that the ENSO phenomenon is the source of much of the variability at these time scales, this indicates that ENSO has been an important part of interannual climatic variations over broad areas of the circum-Pacific region throughout the last millennium. Significant coherence values were found for El Niño and reconstructed Sierra Nevada winter precipitation at $\sim$ 2–4 yr throughout much of their common record (late 1500s to present) and between 6 and 7 yr from the mid-18th to the early 20th century.

At decadal time scales each record generally tends to exhibit significant spectral power over different periods at different times. Both the Quelccaya Ice Cap δ^{18}O series and the Quinn El Niño event record exhibit significant spectral power over frequencies $\sim$ 35 to 45 yr; however, there is low coherence between these two series at those frequencies over their common record. The Sierra Nevada winter rainfall reconstruction exhibits consistently strong variability at periods of $\sim$ 30–60 yr.

1. Introduction

There is considerable interest in expanding our knowledge about the natural variability of climate. Such knowledge can be useful in evaluating whether recent climatic changes are unique in their spatial and temporal characteristics, or possibly reflect anthropogenic forcing. In particular, recent observational and modelling studies of ocean circulation suggest that the oceanic response to perturbations in the climate system can be both large and rather abrupt (Bryan, 1986; Manabe and Stouffer, 1988; Venrick *et al.*, 1987; Dickson *et al.*, 1988; Ebbesmeyer *et al.*, 1991; Gordon *et al.*, 1992). Stocker and Mysak (1992) have evaluated the spectral signatures of several climate indicators and suggest that the characteristic time scales of variability in these records (in the range of 50–400 yr) are indicative of low-frequency ocean-to-atmosphere heat flux variations, and they point to the North Atlantic as a potential major source of this variability.

Increased attention has been focused recently on long-term variability of the El Niño/Southern Oscillation (ENSO) phenomenon (Enfield and Cid, 1991; Quinn and Neal, 1992; Diaz and Pulwarty, 1992; Michaelsen and Thompson, 1992). ENSO represents one of the major modes of natural variability of the earth's climate system; hence, documenting long-term changes in the ENSO cycle in order to understand possible mechanisms associated with such low-frequency variations, would seem useful, not only from the point of view of detecting future anthropogenic changes in climate, but also with the aim of improving climate prediction at decadal time scales.

We have used a suite of long-term indices which are sensitive to some aspect of the ENSO phenomenon. Table I summarizes these indices, their sources, and the period of record covered. All of these indices are located in the general area of influence of the Southern Oscillation (Trenberth and Shea, 1987). One is a direct El Niño chronology developed by W. Quinn and coworkers (Quinn *et al.*, 1987, plus subsequent papers). We have used the latest version of this data set, namely that given in Quinn (1992). Five of the seven series considered here sample portions of the Medieval Warm Period (~ A.D. 950–1250). The other two begin around A.D. 1500, and are included for comparative purposes. A comparison of the temporal variability of these ENSO-sensitive indices may be useful in evaluating how different elements of the large-scale ENSO system, which have only been thoroughly documented for about the last century, may have operated over a much longer period. We are also interested in evaluating whether there are unique elements in these proxy records within the time frame associated with the Medieval Warm Period (MWP), the subject of this special issue.

Our analysis follows two main objectives. One is to describe the characteristic time scales of variability of these ENSO indicators by using the technique known as singular spectrum analysis (Vautard and Ghil, 1989; Ghil and Vautard, 1991; Vautard *et al.*, 1992) in combination with the multitaper method for calculating the spectra of each series (Thomson 1982, 1990a, b; Park *et al.*, 1987). We also

TABLE I: Sources of Proxy ENSO records used in this study. All data are annual, except where noted

	Type of Proxy index	Period of Record
1.	QEN Index. Occurrence of El niño events based on historical and other sources (Quinn *et al.*, 1987; updated in Quinn and Neal, 1992).	A.D. 1525–1985
2.	QNR Index. Compilation of the degree of severity of annual Nile River flood intensity (Quinn, 1992).	629–1520
3.	$Q\delta^{18}O$ Index. Quelccaya Ice-Cap $\delta^{18}O$ values. (Thompson and Mosley-Thompson, 1989).	744–1984
4.	SRNF Index. Reconstructed winter rainfall at Santiago, Chile (Boninsegna, 1988).	1220–1972
5.	PSTMP Index. Reconstructed summer temperatures (from tree-rings) for Rio Alerce, Argentina. (Villalba, 1990).	870–1983
6.	CD/SOI Index. Annual Chinese Drought/SOI Index (Zhang *et al.*, 1989).	1471–1985
7.	SNWR Index. Reconstructed winter precipitation in the southern Sierra Nevada, California. (Graumlich, 1993).	800–1988

evaluate the cospectrum, coherence (coherency-squared) and phase among the different spectra for evidence of temporal association at characteristic ENSO time scales and at the longer time scales. The cross-spectral measures are calculated using multitaper coherence (Thomson, 1982) with jackknifed error estimates as a measure of statistical significance (Thomson and Chave, 1991; Kuo *et al.*, 1990). A second aim is to ascertain what changes, if any, have occurred in the variance spectrum of these different indices over the past several centuries within the ENSO time scale, and at decadal to century scales where oceanic forcing of the atmosphere may be important.

2. Analysis Methods

Two nonparametric procedures, singular spectrum analysis (SSA) and Thomson's (1982) multi-taper method (MTM) are used in combination to analyze the spectrum of the various climate indices. SSA has been shown to be a useful tool for defining the principal models of variability of geophysical series. MTM has been shown to be an improvement over traditional single-window estimates (see Park *et al.*, 1987; and Thomson, 1990a).

SSA, an extension of empirical orthogonal function analysis, considers M lagged copies of a time series process x_i $1 \leq i \leq N$, sampled at equal time intervals τ_s, and calculates the eigenvalues λ_k and eigenvectors ρ_k, $1 \leq k \leq M$ of their covariance matrix $\mathbf{C}$ (see Ghil and Vautard, 1991; Vautard *et al.*, 1992). The matric $\mathbf{C}$ has a Toeplitz structure with constant diagonals corresponding to equal lags. The choice of window width, $\tau_w = M \tau_s$, represents a tradeoff between the amount of information desired to be retained (large M), and statistical significance (small M). Subsets of eigenelements and associated principal components provide for noise reduction, the removal of trend and the identification of oscillatory components (Rasmusson et al., 1990; Penland et al., 1991; Vautard et al., 1992). Selected components provide for optimal reconstruction of particular processes at precise time scales. As in spatial EOF analysis, the complete sum of all eigenelements returns the original input series. In contrast to other standard filtering techniques, where the basis functions are selected a priori, SSA functions as a data-adaptive filter.

The multi-taper method involves multiplication of the input data series by an optimal set of leakage-resistant tapers, yielding several series from one record (see Thomson 1990a, b; Park et al., 1987). A discrete Fourier transform is applied to each of these series producing several 'eigenspectra'. In a single-taper direct estimate of the spectrum of a white noise process at frequency f, and bandwidth $2W$, we aim to maximize Λ, the fraction of spectral energy in that estimate that derives from the frequency interval $|f - f'| \leq W$,

$$\Lambda(N, W) = \frac{\int_{-W}^{W} |A(f)|^2 df}{\int_{-1/2}^{1/2} |A(f)|^2 df}, \tag{1}$$

where $A(f)$ is the discrete Fourier transform of the data tapers, and $1 - \Lambda$ is the fraction of spectral energy that leaks from outside the band. The weighted sum of the eigenspectra form a single spectral estimate with little compromise between leakage and resolved variance. The statistical information discarded by the first taper is partially recovered by the second taper, and so on. Park et al. (1987) and Thomson (1990a) furnish good documentation and applications of the technique.

The tapers minimize loss of information, as is the case with traditional single tapers, while also optimizing resistance to spectral leakage. Many geophysical time series exhibit nonstationary behavior with time. The multitaper method, which is designed to weigh data as evenly as possible from the full record sample, is not subject to problems associated with single taper techniques which discard varying amounts of end data, while overemphasizing signal power over the central portions of the record. The multitaper spectral estimates are a smoothed estimate, and also a consistent estimator of spectral variance. Another advantage of MTM is that the statistical significance of the spectral peaks can be calculated via an F-ratio test.

The evolution of local spectrum in time provides valuable information on the onset, amplification or disappearance of oscillatory behavior. Such behavior is associated with internal mechanisms typical of nonlinear systems (Yiou et al., 1991). This 'evolutive' spectral analysis (ESA) is achieved by selecting a specified window length, calculating the spectrum within each window (via the above method), and repeating the process while moving the time window along through the series being analyzed.

The time series of event-type data – the Quinn El Niño record (QEN) and the Quinn Nile River flood-deficit series (QRN), which are made up of zeros (no event) and ones (event occurrence) – were transformed to a smoother-valued record by converting it to 'events per year' within overlapping 25-yr segments. For example, for a given 25-yr period within the overall period of record, if there were, say, five El Niño events, then the central year of that interval would have a value of 0.20 (5/25 events/yr). This also implies a mean event recurrence interval of 5 yr. To recover information on ENSO time scales, a second series (QENres) was used which consists of the difference between the original (0s and 1s) series and the 'windowed' series. The choice of window length is arbitrary and depends on the analysis objectives. In this case, the high-frequency window contains spectral power from the biennial and ENSO time scale ($\sim$ 2 to 9 yr), as well as time scales associated with solar cycle variability ($\sim$ 11 and 22 yr). This is one of two ENSO-sensitive series that lies entirely outside the MWP (the other being a reconstruction of the Tahiti-Darwin Southern Oscillation Index derived from historical flood/drought records). However, because it is a direct proxy of El Niño conditions in the upwelling regions off the west coast of tropical South America, it was deemed useful to include it for comparison purposes.

Diaz and Pulwarty (1992) analyzed the temporal characteristics of each of the two Quinn records and also compared an overlapping segment (1824–1941) using a 19-yr running window. In that study, we considered the possibility that at least part of the changes reflected in these records may be due to differences in the amount of information available in the historical record. To that end, we analyzed the spectrum of El Niño and Nile River flood-deficit events for different event strength thresholds. The differences between the early and later parts of these records, such as they exist, are not obviously related to historical details of the reconstructions, except possibly a period of more frequent El Niño events that is evident throughout much of the 19th century. Diaz and Pulwarty (1992) also analyzed the modern record of the Southern Oscillation (based on normalized Tahiti minus Darwin sea-level pressure difference since the 1880s) together with a contemporary sequence of El Niño events derived directly from sea surface temperature observations in the eastern tropical Pacific Ocean. Even over this relatively short time period there have been changes in the large-scale structure of ENSO (see Cole *et al.*, 1993). Hence, in the absence of clear evidence that temporal differences in the QEN series are due solely (or mostly) to non-climatic factors, we will proceed here by taking the data at its face value.

We note that only the moderate and stronger event classification have been used in the present study. The question of how to categorize multi-year events (such as the 1939–41 and 1991–93 episodes) is dealt with by following the same approach taken by a number of other investigators (see, Kiladis and Diaz, 1989), namely, that the first of a set of consecutive years with persistent ENSO-like anomalies is taken as year zero of that event. While it is true that other classification approaches would have the effect of changing somewhat the recurrence characteristics of the data (see, Hocquenghem and Ortlieb, 1992), we feel we have taken a consistent approach in this analysis.

The other series used in this study that begins after the nominal end of the MWP is a reconstruction of the Southern Oscillation Index based on a long record of dryness/wetness in eastern China (see Wu and Lough, 1987; Zhang *et al.*, 1989). The influence of the Southern Oscillation on seasonal precipitation in China is substantial, although it is primarily manifested in the subtropical humid regions of eastern China, and is subject to considerable event-to-event variations. It in included here for comparison purposes as representative of an ENSO teleconnection to the monsoon regions of Asia (see Ye *et al.*, 1987).

The SSA was performed on each series using a range of window sizes to check the stability of the eigenvalues. Lag windows of $M = 25$ and $M = 150$ were chosen to isolate components with power in the 2–25 yr and 25–150 yr bands, respectively. A Mann-Kendall test (WMO, 1966) and maximum-entropy spectral analysis (Burg, 1967) were carried out on each principal component to identify probable trends and dominant spectral peak. Separate time-series can be reconstructed using only PCs carrying power in the desired bands. Furthermore, we focus specifically on pairs of adjacent PCs which correspond to oscillatory components. Each reconstructed band series is analyzed via the MTM technique, with bandwidth $W = 5/N$ and $K = 9$ tapers, where N is the length of the respective series. Multi-taper evolutive spectrum of each series is calculated using a moving window of 91 yr with increment of one year. Statistical significance via an F-ratio test is evaluated at each step and can be mapped in the same manner as the evolutive spectrum. Coherence and phase were calculated among all series pairs; however, only cross-spectra from series with significant power in similar spectral bands, and in particular within the high-frequency ENSO band, are discussed in detail below.

3. Data Analysis

3.1. *Characteristic Scales of Variation*

Standardized indices of the Quinn El Niño series (QEN) and the Quinn Nile River flood-deficit series (QNR) that have been SSA-filtered at different wavelengths are shown in Figures 1 and 2, respectively. All series have been standardized by the standard deviation of the original series to facilitate intercomparison. The amount of MTM spectral variance for each record found within the different frequency bands is listed in Table II. The amount of variance for both the transformed 25-yr filtered (QEN25/QNR25) and residual (QENres/QNRres) series are shown separately. The curves illustrate the long-period trends ($P > 150$ yr), and variability within the 25–150 year low-frequency window and in the 2–10 yr time scale. For QEN25, the amount of variance in the 25–150 yr band is 63% compared with 34% for the QNR25 data. Fluctuations at $P > 150$ yr account for three times as much of the corresponding series variance in the Nile River data than for the El Niño (66 versus

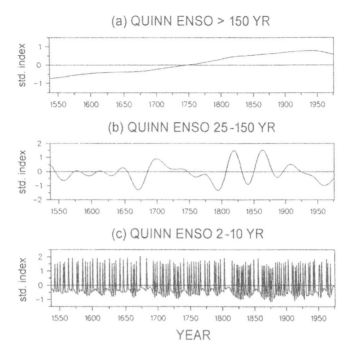

Fig. 1. Band-pass filtered series of the Quinn El Niño (QEN) record (moderate and stronger El Niño occurrence) based on a 25 yr window: (a) low frequency ($P > 150$ yr) component; (b) intermediate ($P \sim 25$–150 yr) time scales; and (c) ENSO time scale ($P \sim 2$–10 yr). Time series in (c) derived from the residual values obtained by differencing the 25-yr running window values obtained from the original data of 0s and 1s. The series were reconstructed by using the appropriate SSA principal components corresponding to the indicated time scale. Ordinate units are in standard deviations of the original series. Data from Quinn (1992).

TABLE II: Percent of variance within indicated frequency bands for each series identified in Table I. Spectrum calculated using the MTM technique on SSA-derived PCs. QEN25 and QNR25 are the transformed Quinn El Niño and Nile River series, respectively, while QENres and QNRres stand for the differenced (original minus QEN25/QNR25) series

	Index	Frequency band			
		> 150 yr	25–150 yr	< 25 yr	2–10 yr
1.	QEN25	21.0	63.0	16.0	–
	QENres	0.0	3.5	96.5	95.0
2.	QNR25	66.0	34.0	0.0	–
	QNRres	0.0	3.0	97.0	78.0
3.	$Q\delta^{18}O$	12.0	17.0	71.0	48.0
4.	SRNF	3.3	9.0	88.0	67.0
5.	PSTMP	2.2	2.2	95.6	72.0
6.	CD/SOI	0.0	8.0	92.0	85.0
7.	SNWR	2.0	7.0	91.0	71.0

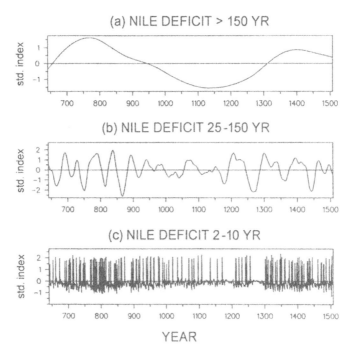

Fig. 2. As in Figure 1, but for the Nile River flood deficit record (QNR) of Quinn (1992). Events in the strongest four categories were used.

21%). In the high frequency band, about 95% of the QEN variability occurs on time scales < 10 yr compared to about 78% for QNR (Table II).

Another notable feature evident in Figures 1 and 2 is a general increase in the frequency of El Niño events over the period of record. To some extent this is likely due to improved sources of information on the South American coast with time. In the absence of other independent records of ENSO activity, such as tropical corals (Cole *et al.*, 1992, 1993), we cannot say for sure if this is entirely an artefact or a real climatic signal. By contrast, the Nile River record shows a period of low variability (few recorded instances of the occurrence of significant deficits) during the MWP, particularly in the 12th century. This can also be noted at intermediate frequencies (Figure 2b), where the amplitude of the oscillations is much reduced during that time.

Michaelsen and Thompson (1992) have shown that the $\delta^{18}O$ record from the Quelccaya Ice Cap in Peru ($Q\delta^{18}O$) has significant variance within the ENSO time scale. The actual data series, together with the low-order trend, and the decadal and higher-frequency filtered series are shown in Figure 3. About 71% of the variance is contained at $P < 25$ yr (see Table II), of which about half is contained on 2–10 yr time scale. The $Q\delta^{18}O$ record displays lower variability and some relatively high values (warmer temperatures) around 1100 A.D. However, the largest signal is associated with a cold period that extends from about the 16th through 19th

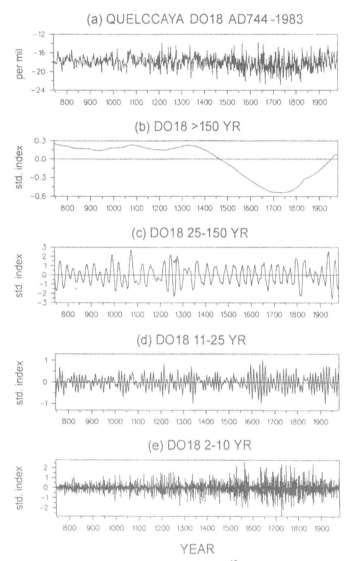

Fig. 3. Original and filtered values of the oxygen isotope ($Q\delta^{18}O$) index for the Quelccaya Ice Cap. Top panel (a) shows the unfiltered data (in units of per mill); panel (b) depicts the lowest frequency trends ($P > 150$ yr); panel (c) depicts variability within the 25–150 time scale; panel (d) shows variability at 10–25 time scales; and panel (e) variability in the 2–10 yr time ENSO band. Data provided by L. Thompson.

centuries, punctuated by greater variability at higher frequencies (Figures 3d and e). This period of sustained negative $\delta^{18}O$ anomalies falls within the time interval commonly referred to as the Little Ice Age (Grove, 1988). About 17% of the variance in the $Q\delta^{18}O$ record is found at intermediate frequencies. Except for a period of relatively high oscillations in the 13th century, we do not see anything notable during the MWP at this time scale. The strongest spectral peak in this band occurs around 30 yr, and explains 2–3% of the total variance in the data.

Two ENSO-sensitive paleoclimate series from South America are reconstructed Santiago de Chile winter rainfall (SRNF) (Boninsegna, 1988), and reconstructed summer temperatures from northern Patagonia in Argentina (PSTMP) (Villalba, 1990). Villalba (this issue, pp. 183–197) has analyzed in detail the association between these two reconstructions and measures of ENSO variability. Here, we will restrict ourselves to an analysis of the characteristic time scales of variability in these two series and whether the association with ENSO, in terms of its main spectral features, has changed over time. Figures 4 and 5 give, respectively, the actual reconstruction, the low-order trends, decadal (25–150 yr), and higher frequency (2–25 yr) variability for SRNF and PSTMP.

The fraction of total variance contained in the ENSO band (2–10 yr) is high – about 67% and 72% for SRNF and PSTMP, respectively. Relatively little variance is found at periods greater than 25 yr (Table II). At least some of the suppression in variance at lower frequencies may be a result of the standardization procedures used in reconstructing the index. During the MWP, reconstructed summer temperatures where at their highest levels compared to the rest of the record (best illustrated in Figure 5b). Variability may also have been higher during these early centuries, although it is difficult to say for sure, since the number of chronologies varies with time, which could affect the variance near the end of such series (see review by Hughes and Diaz, this issue, pp. 109–142). Winter rainfall at Santiago, which is known to be modulated by the ENSO phenomenon (Aceituno, 1988; Kiladis and Diaz, 1989), was lowest during the 14th century, but this was followed by a peak in the 15th century. Although this series starts in A.D. 1220, and hence can not shed any light on possible changes in ENSO frequency during the early part of the MWP proper, the record is useful in that it bridges the gap between the end of the MWP and the start of the El Niño record of Quinn *et al.* (1987). Indeed, the couple of centuries preceding the start of the QEN record appear to have been somewhat unusual.

The SOI reconstruction based on a long record of floods and drought in China (CD/SOI) (Zhang *et al.*, 1989) is presented in Figure 6. Almost all the variance in this series is contained in the 2–10 yr band (Table II). No significant trend was found for this series, so it is likely that trend removal was part of the reconstruction methodology. Significant spectral peaks are found at $\sim$ 3–5 yr, and at $\sim$ 10–11 yr. A variance peak at $\sim$ 75–90 yr is evident in the first half of the record.

The last reconstruction examined consists of a 1000-year record of winter (October-June) precipitation in the Sierra Nevada Mountains of central California (SNWR, Graumlich, 1993). As there is little variance present on time scales > 150 yr (Table II), we show only the series associated with time scales from 2 to 150 yr (Figure 7). There are no notable differences in this record during the MWP compared to other parts of the record, except possibly the presence of some high amplitude variations at frequencies between $\sim$ 13–15 yr and from $\sim$ 35–60 yr.

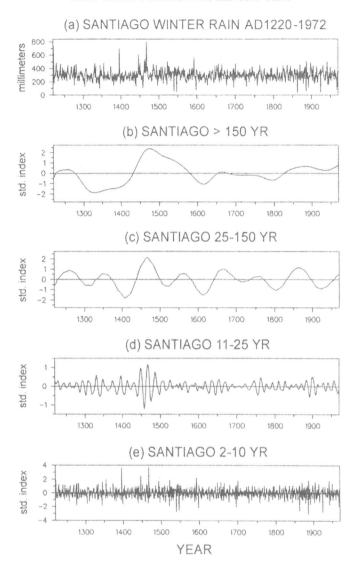

Fig. 4. As in Figure 3, except for reconstructed Santiago de Chile winter rainfall (index SRNF). Values in panel (a) are in mm. Data furnished by R. Villalba.

A separate reconstruction of growing-season temperature (Graumlich, 1993) does indicate the presence of warmer decades during the centuries spanning the MWP (see Hughes and Diaz, this issue, pp. 109–142), as well as cooler conditions from the 16th to the mid-19th century, in agreement with previous tree-ring reconstructions in this region (Briffa *et al.*, 1992). The variations illustrated in Figure 7b exhibit fairly consistent power in the frequency range from 35 to 60 yr. About 30% of the total variance in this band is associated with these time scales. An evolutive spectral analysis plot for this frequency band (Figure 8) illustrates the general strength of

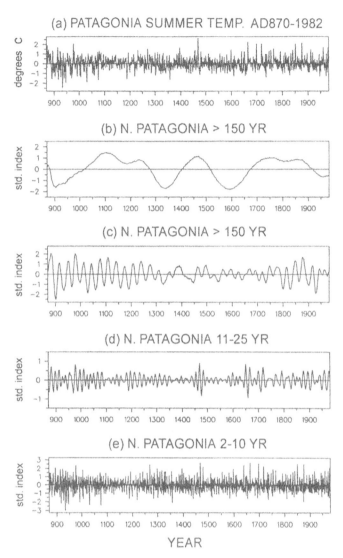

Fig. 5.   As in Figure 3, except for reconstructed northern Patagonia summer temperature (index PSTMP). Values in panel (a) are in degrees Celsius. Data furnished by R. Villalba.

this signal. The high amplitude peaks shown in Figure 7b between about 1000 and 1200 A.D. are associated with oscillations having periods ~ 50–60 yr. After about A.D. 1700, the dominant fluctuations in this band occur at around 35 yr.

We next consider changes with time of the significant spectra within the 2–10 yr frequency band associated with ENSO forcing.

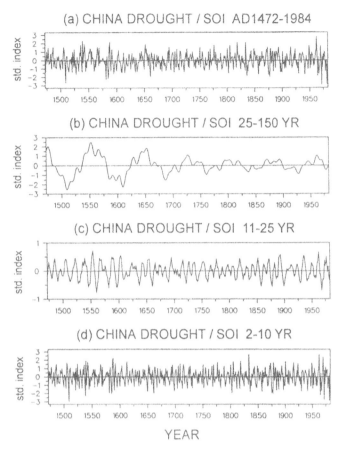

Fig. 6. As in Figure 3, except for reconstructed SOI based on Chinese drought records (index CD/SOI). Shown are (a) the actual data (originally in standardized units); (b) a filtered series tuned to highlight the variability in the 25–150 yr time scale; (c) the variability in the 10–25 yr time scales; and (d) variability in the 2–10 yr ENSO band.

3.2. *Evolutive Spectra*

Figure 9 displays the evolutive spectral signatures of the statistically significant ($\geq 95\%$ confidence level) spectral components for the 2–10 year frequency band for each time series. The pluses, visible in Figure 9 as thicker line segments, denote times when variations at the particular frequency in each evolutive time step exceed 5% of the total variance. Since the emphasis in this study is on the ENSO time scale, and because of space constraints, only the ESA for the high-frequency band (specifically in the 2–10 yr window) is shown. Variance peaks at lower frequencies are discussed as appropriate. We caution the reader that changes in the spectrum of the event data imply a change strictly in the frequency of recurrence, whereas a change in the spectrum of the other series implies a modulation in the time scale of variability in the intrinsic property of the data itself.

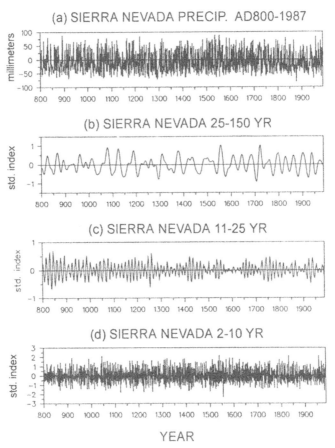

Fig. 7. As in Figure 3, except for reconstructed winter (October-June) precipitation in the Sierra Nevada mountains of central California (index SNWR). Data from Graumlich (1993), furnished by M. Hughes. Shown are (a) the actual data (in mm); (b) a filtered series tuned to highlight the variability at 25–150 yr; (c) 10–25 yr time scale; and (d) variability in the 2–10 yr ENSO band.

As expected, significant power is indicated in the canonical ENSO time scale (∼ 2–6 yr) for the QENres index (Figure 9a). At lower frequency (QEN25 index) spectral peaks occur at ∼ 40–60 yr (evolutive spectrum not shown, see Diaz and Pulwarty, 1992). From the late 1700s to the end of the 19th century, the QEN record exhibits strong variance at ∼ 45 yr (see Figure 1b). Similarly, QNRres index displays significant power between ∼ 2–4 yr throughout most of the record (Figure 9b), but significant low frequency power occurs in different parts of the record for periods of ∼ 40–60 yr and near 90 yr (not shown). We should note that the residual indices, QENres and QNRres, of necessity, must contain a large amount of spectral power at frequencies characteristic of the recurrence time in the original series. However, whether the peak is at the longer ENSO time scale (> 5 yr) or at shorter periods (∼ 3–5 yr) depends on the actual record of El Niño

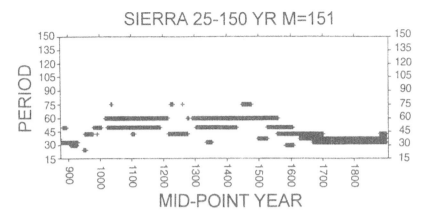

Fig. 8. Evolutive spectral analysis (ESA) for SNWR in the 25–150 yr frequency band.

occurences.

The oxygen isotope record ($Q\delta^{18}O$) exhibits a broader spectral signature with significant power throughout the ENSO band (Figure 9c) and also in the 9–14 yr time scale (see Figure 3). During the MWP, variations in the $Q\delta^{18}O$ record appear to be concentrated at relatively longer periods in the ENSO band ($\sim$ 5–8 yr), whereas later on they occur mostly at somewhat higher frequencies, between $\sim$ 2–6 yr. Significant and continuous power is also found at $\sim$ 25–50 yr (not shown, but see Figure 3c). An increase in negative $\delta^{18}O$ ratios is evident during the Little Ice Age ($\sim$ A.D. 1500–1900). During that time $\delta^{18}O$ variations had a strong rhythm at $\sim$ 30 yr.

The corresponding ESA for SRNF is given in Figure 9d. At higher frequencies, significant and continuous power is evident at $\sim$ 8 yr and between about 2 and 6 yr, and also from $\sim$ 16–22 yr (not shown). Before about A.D. 1500 there are no dominant frequencies at the 2–10 yr time scale, but after about A.D. 1600, variability at $\sim$ 3–4 yr is strong. At the lower frequencies, two harmonics stand out, one at $\sim$ 94 yr, and a second at $\sim$ 125 yr. For northern Patagonia summer temperatures (Figure 9e), consistent variance peaks are found at $\sim$ 3–6 yr, near 10 yr, and a strongly continuous peak is found near $\sim$ 25–40 yr in the 25–150 yr band (not shown).

The principal time scale of variability for the CD/SOI index is concentrated in two narrow bandwidths – between $\sim$ 4–6 yr, and, after 1800, between $\sim$ 7–8 yr as well (Figure 9f). As noted earlier, there is little low frequency variance associated with this index, although a relative peak is found at $\sim$ 70—85 yr periods in the first half of the record (Figure 6b). Lastly, the Sierra Nevada winter precipitation index, exhibits strong variability at $\sim$ 2–5 yr and from $\sim$ 6–8 yr (Figure 9g). It was noted earlier that at lower frequencies, the strongest spectral signatures occur between 35 and 60 yr, with a weaker one near 100 yr.

Table III provides a summary of the principal spectral lines in the 2–10 yr

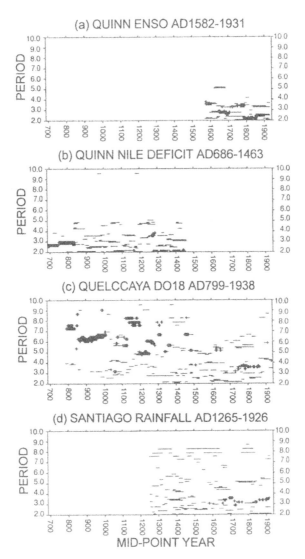

Fig. 9.   Evolutive spectral analysis (ESA) for different indices: (a) QENres; (b) QNRres; (c)Qδ^{18}O;
(d) SRNF; (e)PSTMP; (f)CD/SOI; (g) SNWR in the 2–10 yr spectral band. Two symbols are used, a
straight line for values that account for less than 5% of the total variance in the data and pluses (+)
to denote 91-yr segments when the line spectra account for more than 5% of the data variance in that
particular segment. Only components significant at the 95% confidence level are shown.

frequency band (column labelled LINES). It also lists the periods where significant
coherence peaks are found among the different series. The coherence features are
discussed further below. The significant spectral peaks occur at frequencies near
the tropical quasi-biennial oscillation (QBO), near the mean recurrence frequency
for ENSO events (3–4 yr) and at periods in the neighborhood of 6 yr.

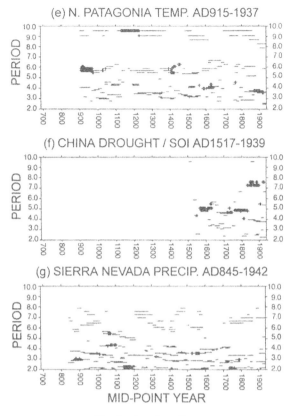

Fig. 9. (continued)

To further illustrate the persistence of the major periods found in these long-term ENSO indices, Figure 10 illustrates the fraction of time where a significant spectral peak has been plotted in Figure 9. For comparison, the corresponding fractions of time where significant spectral peaks are found within the MWP are included. Only the five series containing data in the MWP are shown. The Nile (QNRres) and Sierra Nevada (SNWR) indices (Figure 10a and e) display consistent oscillations in the 2–4 yr time scale throughout their respective periods of record. The $Q\delta^{18}O$ series exhibits relative low percentages before ~ 5 yr, but consistent variability from about 5 to 9 yr (Figure 10b). Northern Patagonia summer temperature (PSTMP) exhibits consistent variance peaks around 5.7 and 9.7 yr, while the Santiago, Chile reconstruction (SRNF) exhibits a persistent variance peak around 8.1 yr (Figure 10c and d).

3.3. *Coherence Spectra*

One purpose of this study was to ascertain the degree of coherence among the various series within the ENSO time scale, and the stability of these relationships

TABLE III: Oscillatory components obtained via SSA (LINES col.) and significant coherence spectra at the indicated harmonics which correspond to periods
where both series have significant variance spectrum components at the 90 and
95% (bold figures) significance levels

	LINES	$Q\delta^{18}O$	SRNF	PSTMP	CD/SOI	SNWR
QENres	**2.2**, **2.8**, **3.3**, **6.7**	**2.2**, **5.7**	3.3, **4.5** 6.6	**2.2**, **3.3** 6.1	**2.8**, **5.9**	**2.2**, 2.9, 6.7
QNRres	**2.2**, **2.9**, 3.6	**2.4**, 3.6	3.8,	**2.4**, 3.4, 8.1		**2.2**, 6.7
$Q\delta^{18}O$	**2.2**, **3.6**, 6.4		**3.8**, **3.4** 4.1, 8.1	**2.2**, **4.4** 6.5	**3.4**, **4.1** 9.1	**2.2**, 3.1, 5.9, **7.5**
SRNF	2.9, **3.4**, **8.1**			**2.4**, **6.1** 4.8	**3.4**, **4.1**	2.6, **3.4**, 7.4
PSTMP	**3.8**, **5.7**, **8.5**, **9.7**				**2.2**, 3.5 **4.2**	2.7, 4.8
CD/SOI	**2.4**, **4.9**, **7.5**					**2.4**, 3.4 **5.4**
SNWR	**2.2**, **3.6**, **6.7**					

over time. We have calculated coherence spectra using the method outlined in
Thomson (1982) and Thomson and Chave (1991), and have applied that method
in an evolutive fashion as described earlier. Figure 11 shows squared coherency
spectra in the 2–10 yr ENSO band for selected pairs of indices. Panels 11a–d show
the coherency-squared and phase between the QENres and the $Q\delta^{18}O$, SRNF,
PSTMP, and SNWR, indices, respectively. The Quinn El Niño record was used as
the reference series, since this is the only one of the seven series used here that
can be considered to be a 'true' index of ENSO activity (the other locations being
forced by ENSO variability through teleconnection processes). Although the QEN
series, unfortunately does not extend back to the MWP, we feel that some insight
can be gained by comparing the behavior of the different series through a span of
more than four centuries.

The highest coherence values between the QENres and $Q\delta^{18}O$ series (all significant at between 95 and 99% confidence levels) occur in the frequency band
from 5 to 6 yr (Figure 11a), with the QENres series leading the $Q\delta^{18}O$ by 2 to 3
months at ~ 5.7 yr (Table III). Typically, oxygen isotopic ratios are less negative
(indicative of warmer prevailing conditions) in association with major warm events
of the Southern Oscillation. At the same time, the accumulation record indicates
lower precipitation values during El Niño years. The peak in the annual cycle of
precipitation in this region of the Peruvian Andes occurs during northern spring.
The reader is referred to Thompson and Mosley-Thompson (1989) for particulars
of the ice core record on the Quelccaya Ice Cap.

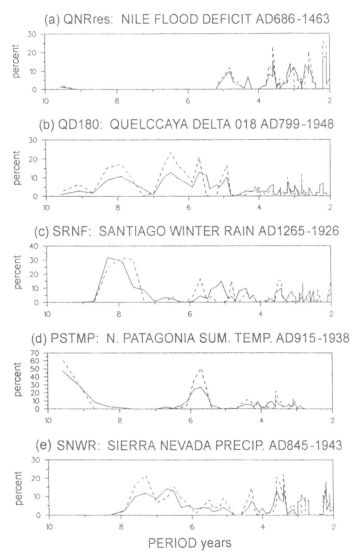

Fig. 10. Percentage of time that a significant line spectra is found in the evolutive spectrum of (a) QENres; (b) $Q\delta^{18}O$; (c) SRNF; (d)PSTMP; and (e) SNWR over the full period of record (solid line) and within the MWP (A.D. 900–1400, dashed line).

By and large, we find strong coherence between the Quinn El Niño (QENres) series and the other series within the ENSO time scale of ∼ 2–6 yr. The coherence spectra between QENres and SRNF is also interesting. In the statistically significant band near 4 yr (Figure 11b), SRNF leads QENres by about 4 months, which is roughly what is observed today, as winter (June-August) rainfall in Santiago de Chile tends to be above normal during a developing El Niño, several months before the event reaches a peak along the Peruvian coast (Kiladis and Diaz, 1989).

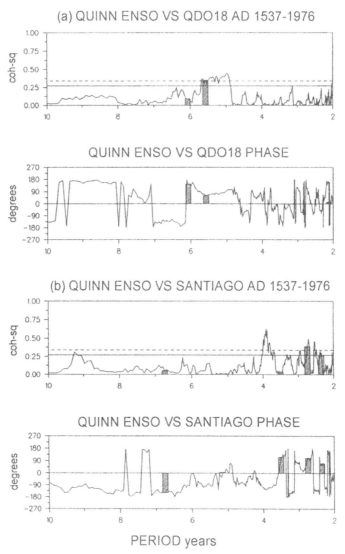

Fig. 11. Coherency-squared and phase calculated for the 2–10 yr ENSO frequency band. Solid and dashed horizontal lines denote 95% and 99% significance levels, respectively. Panel (a) QENres vs $Q\delta^{18}O$; (b) QENres vs SRNF; (c) QENres vs PSTMP; and QENres vs CD/SOI. Vertical shaded bars highlight frequencies that exhibit oscillatory components in the autospectra of both series (see Table III). The height of the bar corresponds to the coherency and phase value at the particular point in the respective plot.

The coherence spectra for QENres and Patagonia summer (December-April) temperature reconstruction (PSTMP, Figure 11c) exhibit major peaks at 3.3, 4, and 6 yr, with the El Niño index leading by about four months in each instance. Three significant coherency peaks, in the frequency interval from ~ 2 to 4 yr periods, are evident between QENres and SNWR (Figure 11d). The phase is different for each

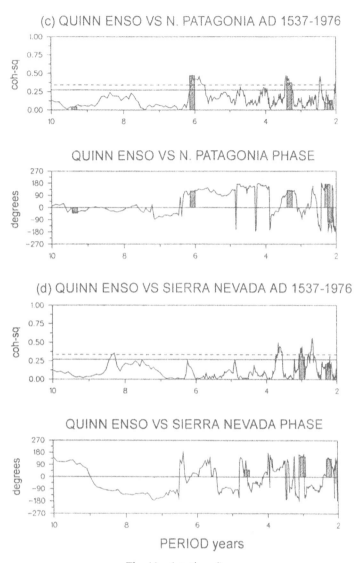

Fig. 11. (continued)

peak, but for the one around 3.7 yr, which is close to the long-term mean El Niño recurrence in the modern record, QNRres leads SNWR by about two months.

Highly statistically significant coherence peaks between QENres and CD/SOI are found between 2 and 3 yr, and around 6 yr (figure not shown), with QNRres leading by about 3 months. High coherence is found between CD/SOI and SRNF in the ~ 2–4 yr band (figure not shown), but only the peaks between 3 and 4 yr are consistent with an ENSO-type signal, as SRNF is below normal when the reconstructed SOI is high (La Niña conditions) and viceversa for low SOI values (El Niño conditions). Indeed, what these analyses show is that the ENSO signal,

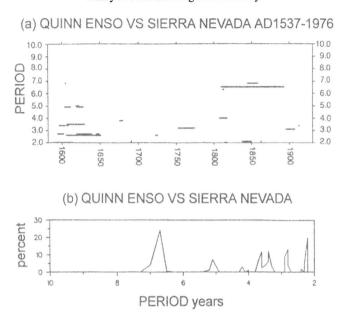

Fig. 12. Evolutive coherence between QENres and SNWR, (a) coherence-squared at 90% significance level, with the stipulation that both individual peaks must also be significant at the 90% level; (b) percentage of time that a significant coherency value is plotted in (a).

as reflected in variations in the frequency of recurrence of the event data, and in amplitude/frequency variations in the continuous records, is reasonably well manifested in the coherence spectra among the series.

We calculated the evolutionary coherence spectrum for the different series pairs within the 2–10 yr ENSO frequency band. However, unlike the ESA of the autospectra, the persistence of the significant coherence peaks through time is considerably lower. In general, significant peaks occur only over short time intervals. One of the most consistent cospectra is between QENres and SNWR (Figure 12). Here, ENSO covariability is fairly well preserved around 3–4 and 6–7 yr. The lower panel shows the percent of time the coherency is significant in the 91–yr running window at the 90% confidence level.

4. Summary and Discussion

The band-specific behavior of a suite of long period temperature and precipitation indices were examined. These indices covered a broad zone around the Pacific centers of action of the Southern Oscillation and have previously been shown to be sensitive indicators of climatic effects associated with the changing phases of the SO. Our aim was to describe the basic temporal characteristics of these ENSO-sensitive time series, and to ascertain if the variability in these records on time scales pertinent to the ENSO phenomenon was consistent throughout the past several centuries. To accomplish these goals, we have used the techniques

TABLE IV: Summary listing of significant spectral peaks in the SSA/filtered time series described in Table I.

	Index	High-frequency peaks	Low-frequency peaks
1.	QEN	~2–6 yr	~40–60 yr
2.	QNR	~2–5 yr	~40–60 yr; ~90 yr
3.	$Q\delta^{18}O$	~2–8 yr; 9–14 yr	~25–50 yr
4.	SRNF	~8 yr; 16–22 yr	~94 & 125 yr
5.	PSTMP	~3–6 yr; ~10 yr	not significant
6.	CD/SOI	~3–8 yr	~70–85 yr
7.	SNWR	~2–8 yr	~35–60 & 100 yr

of singular spectrum analysis (SSA) in combination with the multitaper methods (MTM) of spectral estimation.

Table IV summarizes the principal time scales of variability in the seven ENSO-sensitive indices analyzed in this study. For nearly all the indices, we found consistent, significant spectral peaks in the frequency band from $\sim$ 2–6 yr, which encompasses both the biennial and ENSO time scales. Spectral coherence between the indices at these frequencies was generally significant. Temporally coherent spectral power at intermediate frequencies ($\sim$ 25–150 yr) was also found, although coherence among the various indices at this frequency range was typically low. A few of the series also exhibited very low frequency (> 150 yr) behavior whose authenticity is unknown. Within the time interval encompassed by the Medieval Warm Period (here broadly defined as $\sim$ 900–1400 A.D.), only the Nile River flood-deficit record and the northern Patagonia summer temperature index appear to show extraordinary behavior. In the former case, much lower incidences of Nile flood deficits are recorded, whereas for the latter, warmer summers are indicated. The Sierra Nevada record exhibits a period of large precipitation fluctuations lasting about one to two centuries. The summer temperature record derived from tree-ring data at the same site does indicate warmer summer during the MWP (Graumlich, 1993, see also Hughes and Diaz, this issue, pp. 109–142). The $\delta^{18}O$ record also exhibits a marked dip (cooler temperatures) during the core of the Little Ice Age.

It appears that the climatic variability expressed at the ENSO time scale is retained quite consistently in nearly all of these climate-sensitive records (of course, the QEN series is implictly a record of El Niño occurrences on the South America Pacific coast). Low frequency behavior of the QEN series has been previously documented (Enfield and Cid, 1991; Diaz and Pulwarty, 1992), and additional high-resolution (annual or seasonal) ENSO-sensitive indices are becoming available for studying low frequency variations in the ENSO system (Cole *et al.*, 1992, 1993). Given the limited sampling available during the MWP, the results of our study suggest that there were not major differences in the climatic expression of the ENSO phenomenon during the last millenia, a period encompassing a warm epoch (at least regionally, see a review by Hughes and Diaz, this issue, pp. 109–142) and a generally colder period (the 'Little Ice Age'). Enfield (1992) and Nicholls (1992), on the basis of various lines of evidence, both argue for a relatively stable ENSO system. Although modulations in the relationship between El Niño and the Southern Oscillation have been documented for about the past century (Trenberth and Shea, 1987; Deser and Wallace, 1987; Cole *et al.*, 1993), and other studies suggest differences over longer periods (see Diaz and Markgraf, 1992), it appears that ENSO may have operated in a substantially similar fashion over the past thousand years as it has during the past century.

Acknowledgements

We wish to thank Lisa Graumlich and Ricardo Villalba for making their tree-ring reconstructions available to us. Villalba also kindly provided Boninsegna's reconstruction for Santiago de Chile rainfall (that record is discussed in a book edited by Bradley and Jones (1992). Records for the other sites were obtained from the published sources. We also thank the two reviewers for their valuable comments which helped to improve the manuscript.

References

Aceituno, P.: 1988, 'On the Functioning of the Southern Oscillation in the South American Sector', *Mon. Wea. Rev.* **116**, 505–524.

Boninsegna, J. A.: 1988, 'Santiago de Chile Winter Rainfall since 1220 as Being Reconstructed by Tree-Rings', *Quatern. South Amer. Antarc. Penins.* **6**, 67–87.

Bradley, R. S. and Jones, P. D. (eds.): 1992b, *Climate Since A.D. 1500,* Routledge, London and New York, 679 pp.

Briffa, K. R., Jones, P. D., and Schweingruber, F. H.: 1992, 'Tree-Ring Density Reconstructions of Summer Temperature across Western North America', *J. Clim.* **5**, 735–754.

Bryan, F.: 1986, 'High-Latitude Salinity Effects and Interhemispheric Thermohaline Circulation', *Nature* **323**, 301–304.

Burg, J. P.: 1967, 'Maximum entropy spectral analysis', Reprinted in *Modern Spectrum Analysis* (1978), D. G. Childers (ed.), IEEE Press, NY, pp. 42–48.

Cole, J. E., Shen, G. T., Fairbanks, R. G., and Moore, M.: 1992, 'Coral Monitors of El Niño/Southern Oscillation Dynamics across the Equatorial Pacific', in Diaz, H. F. and Markgraf, V. (eds.), *El Niño: Historical and Paleoclimatic Aspects of the Southern Oscillation*, Cambridge University Press, Cambridge, pp. 349–375.

Cole, J. E. Fairbanks, R. G., and Shen, G. T.: 1993, 'Recent Variability in the Southern Oscillation: Isotopic Results from a Tarawa Atoll Coral', *Science* **260**, 1790–1793.

Deser, C. and Wallace, J. M.: 1987, 'El Niño Events and Their Relation to the Southern Oscillation: 1925–1986', *J. Geophys. Res.* **92**, 14,189–14,196.

Diaz, H. F. and Markgraf, V.: 1992, *El Niño: Historical and Paleoclimatic Aspects of the Southern Oscillation*, Cambridge University Press, Cambridge, 476 pp.

Diaz, H. F. and Pulwarty, R. S.: 1992, 'A Comparison of Southern Oscillation and El Niño Signals in the Tropics', in Diaz, H. F. and Marktraf, V. (eds.), *El Niño: Historical and Paleoclimatic Aspects of the Southern Oscillation*, Cambridge Univesity Press, Cambridge, pp. 175–192.

Dickson, R. R., Meincke, J., Malmberg, S. -A., and Lee, A. J.: 1988, 'The "Great Salinity Anomaly" in the Northern North Atlantic 1968–1982', *Prog. Oceanogr.* **20**, 103–151.

Ebbesmeyer, C. C., Cayan, D. R., McLain, D. R., Nichols, F. H., Peterson, D. H., and Redmond, K. T.: 1991, '1976 Step in the Pacific Climate: Forty Environmental Changes between 1968–1975 and 19770–1984', *Proc. Seventh Ann. Pacific Climate (PACLIM) Workshop*, California Dept. of Water Resources.

Enfield, D. B.: 1992, 'Historical and Prehistorical Overview of the El Niño/Southern Oscillation', in Diaz, H. F. and Markgraf, V. (eds.), *El Niño: Historical and Paleoclimatic Aspects of the Southern Oscillation*, Cambridge University Press, Cambridge, pp. 95–117.

Enfield, D. B. and Cid S., L.: 1991, 'Low-Frequency Changes in El Niño-Southern Oscillation', *J. Clim.* **4**, 1137–1146.

Ghil, M. and Vautard, R.: 1991, 'Interdecadal Oscillations and the Warming Trend in Global Temperature Time Series', *Nature* **350**, 324–327.

Gordon, A. L., Zebiak, S. E., and Bryan, K.: 1992, 'Climate Variability and the Atlantic Ocean', *EOS* **73**, 161–165.

Graumlich, L. J.: 1993, 'A 1000-Year Record of Temperature and Precipitation in the Sierra Nevada', *Quatern. Res.* **39**, 249–255.

Grove, J. M.: 1988, *The Little Ice Age*, Methuen, London and New York.

Hocquenghem, A. -M. and Ortlieb, L.: 1992, 'Historical Record of El Niño Events in Peru (XVI–XVIIIth Centuries): The Quinn *et al.* (1987) Chronology Revisited', in Ortlieb, L. and Macharé, J. (eds.), *Paleo-ENSO Records international symposium* (Extended abstracts), ORSTOM-CONCYTEC, Lima, Peru, pp. 143–149.

Kiladis, G. N. and Diaz, H. F.: 1989, 'Global Climatic Anomalies Associated with Extremes of the Southern Oscillation', *J. Clim.* **2**, 1069–1090.

Kuo, C., Lindberg, C., and Thomson, D. J.; 1990, 'Coherence Established between Atmospheric Carbon Dioxide and Global Temperature', *Nature* **343**, 709–714.

Manabe, S. and Stouffer, R. J.: 1988, 'Two Stable Equilibria of a Coupled Ocean-Atmosphere model', *J. Clim.* **1**, 841–866.

Michaelsen, J. and Thompson, L. G.: 1992, 'A Comparison of Proxy Records of El Niño/Southern Oscillation', in Diaz, H. F. and Markgraf, V. (eds.), *El Niño: Historical and Paleoclimatic Aspects of the Southern Oscillation*, Cambridge University Press, Cambridge, pp. 323–348.

Nicholls, N.: 1992, 'Historical El Niño/Southern Oscillation Variability in the Australasian Region', in Diaz, H. F. and Markgraf, V. (eds.), *El Niño: Historical and Paleoclimatic Aspects of the Southern Oscillation*, Cambridge University Press, Cambridge, pp. 151–173.

Park, J., Lindberg, C. R., and Vernon III, F. L.: 1987, 'Multitaper Spectral Analysis of High-Frequency Seismograms', *J. Geophys. Res.* **92**, 12,675–12,684.

Penland, C., Ghil, M., and Weickmann, K.: 1991, 'Adaptive Filtering and Maximum Entropy Spectra with Application to Changes in Atmospheric Angular Momentum', *J. Geophys. Res.* **96**, 22,659–22,671.

Quinn, W.H.: 1992, 'A Study of Southern Oscillation-Related Climatic Activity for A.D. 622–1990', in Diaz, H. F. and Markgraf, V. (eds.), *El Niño: Historical and Paleoclimatic Aspects of the Southern Oscillation*, Cambridge University Press, Cambridge, pp. 119–149.

Quinn, W. H., Neal, V. T., and Antunez de Mayolo, S. E.: 1987, 'El Niño over the Past Four and Half Centuries', *J. Geophys. Res.* **92**, 14449–14461.

Quinn, W. H. and Neal, W. T.: 1992, 'The Historical Record of El Niño Events', in Bradley, R. S. and Jones, P. D. (eds.), *Climate Since A.D. 1500*, Routledge, London, pp. 623–648.

Rasmusson, E. M., Wong, X. and Ropelewski, C. E.: 1990, 'The Biennial Component of ENSO Variability', *Journal of Marine Systems* **1**, 71–96.

Stocker, T. F. and Mysak, L. A.: 1992, 'Climatic Fluctuations on the Century Time Scale: A Review of High-Resolution Proxy Data and Possible Mechanisms', *Clim. Change* **20**, 227–250.

Thompson, L. G. and Mosley-Thompson, E.: 1989, 'One-Half Millennia of Tropical Climate Variability as Recorded in the Stratigraphy of the Quelccaya Ice Cap, Peru', in Peterson, D. H. (ed.), *Aspects of Climate Variability in the Pacific and the Western Americas*, Geophysical Monograph 55, Amer. Geo. Union, Washington, D.C., pp. 15–31.

Thomson, D. J.: 1982, 'Spectrum Estimation and Harmonic Analysis', *Proc. IEEE* **70**, 1055–1096.

Thomson, D. J.: 1990a, 'Time Series Analysis of Holocene Climate Data', *Phil. Trans. R. Soc. Lond.* **330**, 601–616.

Thomson, D. J.: 1990b, 'Quadratic-Inverse Spectrum Estimates: Applications to Paleoclimatology', *Phil. Trans. R. Soc. Lond.* **332**, 539–597.

Thomson, D. J. and Chave, A. D.: 1991, 'Jackknifed Error Estimates for Spectra, Coherences, and Transfer Functions', in Haykin, S. (ed.), *Advances in Spectrum Analysis and Array Processing*, Vol. I, pp. 58–113.

Trenberth, K. E. and Shea, D. J.: 1987, 'On the Evolution of the Southern Oscillation', *Mon. Wea. Rev.* **115**, 3078–3096.

Vautard, R. and Ghil, M.: 1989, 'Singular Spectrum Analysis in Nonlinear Dynamics with Applications to Paleoclimatic Time Series', *Physica D*, **35**, 395–424.

Vautard, R., Yiou, P., and Ghil, M.: 1992, 'Singular-Spectrum Analysis: A Toolkit for Short Noisy Chaotic Signals', *Physica D* **38**, 95–126.

Venrick, E. L., McGowan, J. A., Cayan, D. R., and Hayward, T. L.: 1987, 'Climate and Chlorophyll a: Long-Term Trends in the Central North Pacific Ocean', *Science* **238**, 70–72.

Villalba, R.: 1990, 'Climatic Fluctuations in Northern Patagonia during the Last 1000 Years as Inferred from Tree-Ring Records', *Quat. Res.* **34**, 346–360.

World Meteorological Organization: 1966, *Climatic Change*, WMO Tech. Note 79, 77 pp.

Wu, X. and Lough, J. M.: 1987, 'Estimating North Pacific Summer Sea Level Pressure back to 1600 Using Proxy Records from China and North America', *Adv. Atmos. Sci.* **4**, 74–84.

Ye, D., Fu, C., Chao, J., and Yoshino, M. (eds.): 1987, *The Climate of China and Global Climate*, Springer-Verlag, 441 pp.

Yiou, P., Genthon, C., Ghil, M., Jouzel, J., Le Treut, H., Barnola, J. M., Lorius, C., and Korotkevitch, Y. N.: 1991, 'High-Frequency Paleovariability in Climate and CO_2 Levels from Vostok Ice Core Records', *J. Geophys. Res.* **96**, 20, 365–20, 378.

Zhang, X., Song, J., and Zhao, Z.: 1989, 'The Southern Oscillation Reconstruction and Drought/Flood in China', *Acta Meteorol. Sinica.* **3**, 290–301.

(Received 19 January, 1993; in revised form 25 October, 1993).

CPSIA information can be obtained
at www.ICGtesting.com
Printed in the USA
LVOW04s2353200116

471563LV00003B/38/P